Collins

Edexcel GCSE (9-1)
Maths for Post-16

Fiona Mapp with Su Nicholson

Jayne Roper and Paul Winters

HarperCollins
PUBLISHERS

Since 1817

William Collins' dream of knowledge for all began with the publication of his first book in 1819. A self-educated mill worker, he not only enriched millions of lives, but also founded a flourishing publishing house. Today, staying true to this spirit, Collins books are packed with inspiration, innovation and practical expertise. They place you at the centre of a world of possibility and give you exactly what you need to explore it.

Collins. Freedom to teach

An imprint of HarperCollins*Publishers*
The News Building
1 London Bridge Street
London SE1 9GF

> Browse the complete Collins catalogue at
> www.collins.co.uk

MIX
Paper from
responsible sources
FSC www.fsc.org **FSC™ C007454**

FSC™ is a non-profit international organisation established to promote the responsible management of the world's forests. Products carrying the FSC label are independently certified to assure consumers that they come from forests that are managed to meet the social, economic and ecological needs of present and future generations, and other controlled sources.

Find out more about HarperCollins and the environment at
www.harpercollins.co.uk/green

Text by Fiona Mapp
Introduction and exam tips by Su Nicholson
Practice and exam-style questions by Jayne Roper and Paul Winters
Commissioned by Jennifer Hall
Project edited by Alexander Rutherford
Development edited by Susan Gardner
Project managed by Maheswari PonSaravanan of Jouve India
Copyedited by Marie Taylor
Proofread by Peter Batty
Answer checked by Steven Matchett
Designed by 2hoots publishing services Ltd.
Typeset by Jouve India
Illustrated by Ann Paganuzzi
Cover designed by ink-tank and associates

Printed and bound by Grafica Veneta S.p.A

The publishers wish to thank the following for permission to reproduce photographs. Every effort has been made to trace copyright holders and to obtain their permission for the use of copyright materials. The publishers will gladly receive any information enabling them to rectify any error or omission at the first opportunity.
(t = top, c = centre, b = bottom, r = right, l = left)

Cover: Crystal Home/Shutterstock, Gordon Saunders/Shutterstock, StudioSmart/Shutterstock, Max Krasnov/Shutterstock p10r Claudio Baldini/Shutterstock, p10l yurchello108/Shutterstock, p11t Jouke van Keulen/Shutterstock, p11b MyImages - Micha/Shutterstock, p12r Iurii Kachkovskyi/Shutterstock, p12l Hongqi Zhang/Alamy, p16 boreala/Shutterstock, p53 Professional Foto CL/Shutterstock, p135 Andy Dean Photography/Shutterstock, p154a Hong Vo/Shutterstock, p154b Denis Larkin/Shutterstock, p154c donatas1205/Shutterstock, p154d bouybin/Shutterstock, p154e holbox/Shutterstock, p154f ever/Shutterstock, p154g Saigonese Photographer/Shutterstock, p154h Alter-ego/Shutterstock, p154i MR.Pongsiam Khumsoithong/Shutterstock, p154j Fotocrisis/Shutterstock, p314 Marcos Mesa Sam Wordley/Shutterstock

With thanks to the students and teachers at the following schools and colleges for their help with this project: Chichester College; City of Bristol College; City Lit College; King Edward VII School, Sheffield; Loughborough College; Milton Keynes College; Nunthorpe Academy; Oaklands College; Peterborough Regional College; The Sixth Form College, Farnborough; West Suffolk College; Whickham School & Sports College. Especial thanks to Anna Bellamy for sharing her PhD research on success in GCSE Maths resits, Catherine Sezen (head of 14–19 at the Association of Colleges) for help and advice, and to Graham Cumming, Edexcel's Maths subject advisor, for his advice and contribution.

Contents

Introduction

Why bother with GCSE Maths?

A qualification in mathematics can provide your passport for progression throughout your education and future career. Due to the high demand for jobs and the number of applications employers receive, they are able to discriminate between applications and many will request a GCSE maths and English qualification of at least grade C, which has become the new grade 4 in Maths GCSE (9-1).

Employers recognise that a mathematics GCSE qualification indicates that the applicant will have the basic number and problem-solving skills which can be a valuable asset in the workplace. Employees with these qualifications are also more likely to be considered for promotion and hence you are likely to receive higher salaries than fellow employees without a maths GCSE qualification.

In higher education, a maths GCSE qualification at grade C/4 is a pre-requisite for many degree courses. As in employment, many courses have far more applicants than they have places, so they set an entry requirement that will allow them to reduce the number of students to consider.

In terms of career choice, you need at least a grade C/4 in mathematics GCSE for the professions of social work, nursing, secondary school and primary school teaching. All things considered, you owe it to yourself to do your best to try and achieve a mathematics GCSE qualification. It can make a big difference for you.

What you need to do to succeed

If you are reading this it is highly likely that you are now re-sitting mathematics GCSE at a Sixth Form college, FE college or evening class. Students in your position often feel despondent. For many of you mathematics may be a subject you didn't really enjoy at school and is the last subject you wanted to study again in the next stage of your education. First of all, you need to get over your feeling of failure. It is OK to fail; we all learn from failure. Think about what you did right and what you need to do to improve. Anything that is worth having is worth working for. It is now in your hands to get focussed, work hard and make sure you take advantage of all the support available to achieve your goal.

This book can be used to support your mathematics classes and ensure you identify the gaps in your knowledge so you know which areas you need to work on to succeed.

Read through the information on the next few pages which outlines how you should prepare for your maths GCSE re-sit. It takes you through the structure of the exams and how to tackle different types of question with pointers to guide you to success. Now it's time to believe in yourself and take control of your future.

GOOD LUCK!

How your Edexcel GCSE (9-1) Maths exam is structured

The Edexcel GCSE (9-1) in Mathematics is assessed through **three examination papers** either at Foundation tier or Higher tier.

The papers are all equally weighted, which means each paper is worth $33\frac{1}{3}\%$ of the final mark.

Foundation (grades 1 – 5)	Paper 1	Paper 2	Paper 3
	Non-calculator 1 hour 30 minutes 80 marks	Calculator 1 hour 30 minutes 80 marks	Calculator 1 hour 30 minutes 80 marks
Higher (grades 4 – 9)	**Paper 1**	**Paper 2**	**Paper 3**
	Non-calculator 1 hour 30 minutes 80 marks	Calculator 1 hour 30 minutes 80 marks	Calculator 1 hour 30 minutes 80 marks

Each paper is worth **80 marks** and the time allowance is **1 hour 30 minutes**, which means you need to allow just over 1 minute for each mark, (8 marks is 9 minutes).

The **instructions** on the front page of each paper state the equipment you need to have for each of the three papers:

• Ruler graduated in centimetres and millimetres
• protractor
• pair of compasses
• pen
• HB pencil
• eraser.

In addition, you will need a **calculator** for papers 2 and 3.

Edexcel examination papers will assess the mathematical content in these proportions:

Mathematical Content	Foundation tier	Higher tier
Number	25%	15%
Algebra	20%	30%
Ratio, proportion and rates of change	25%	20%
Geometry and measures	15%	20%
Probability and Statistics	15%	15%

The chapters in this book are colour-coded to show you which subject areas they cover.

There are three **Assessment Objectives** for the examination papers.

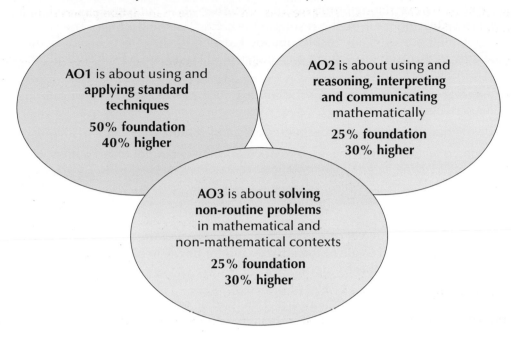

AO1 is about using and **applying standard techniques**

50% foundation
40% higher

AO2 is about using and **reasoning, interpreting and communicating** mathematically

25% foundation
30% higher

AO3 is about **solving non-routine problems** in mathematical and non-mathematical contexts

25% foundation
30% higher

AO1 questions

These questions involve accurately recalling facts, terminology and definitions and accurately carrying out routine procedures. Examples include straight-forward questions on:

- working with fractions and percentages;
- solving equations and inequalities;
- using scale factors;
- plotting given points on a graph;
- using Pythagoras and trigonometry in right-angled triangles.

AO2 questions

These questions involve making deductions and drawing conclusions from mathematical information and presenting arguments and proofs. Examples include:

- using familiar angle properties in more complex problems;
- interpreting statistical diagrams e.g. trends and correlation;
- interpreting timetables;
- interpreting information to perform calculations;
- using ratio in more complex problems;
- criticising questions in questionnaires;
- using experimental probability.

AO3 questions

These questions involve translating problems in non-mathematical contexts into a series of mathematical processes and interpreting results in the context of the problem. They could also involve evaluating methods used and results obtained to identify the effects of any assumptions made. The majority of the problem-solving questions are AO3 and can be solved using more than one method. Examples include:

- assessing numeracy or measure in a realistic context, such as household bills or catering, where decisions need to be made;
- problem-solving questions involving perimeter, area and volume of 2D and 3D shapes;
- problem-solving questions involving, e.g. connections between numbers or angles in a polygon, which can be solved using algebraic techniques;
- geometric reasoning questions where angles need to be found and reasons given for each stage of working.

In different exam questions, you will gain marks not just for getting the right answer but for the method or process you've used to get to the answer, and for how you have communicated the answer. The mark schemes use these letters:

M	method mark awarded for a correct method or partial method
P	process mark awarded for a correct process as part of a problem-solving question
A	accuracy mark awarded after a correct method or process
C	communication mark
B	unconditional accuracy mark

Always remember to show your workings as clearly and fully as possible – especially on a question with several marks – to make sure that you gain all the marks possible.

Understanding exam language

These are some of the common commands used in exam questions:

Question Command	What this means
Estimate	Commonly means round numbers to 1 significant figure to work out a calculation
Calculate	Some working out is needed so it should be written down
Work out or find	A written or mental calculation is needed
State or write down	A written calculation is not needed, could be reading a value from a graph or table
Give the exact value of	Do not round or approximate your answer e.g. $\frac{1}{3}$ must not be written as 0.33. Phrase can be used when answer is wanted in surd form or in terms of π
Solve algebraically	Commonly used for solving simultaneous equations, means algebra must be used; other methods, such as trial and improvement, will not gain marks
Hence work out or find	Used in part questions, means use what you have found in the previous part to find the solution to a question

Hence or otherwise work out or find	Means you can use what you have found in the previous part or another method, to find the solution to a question
Give your answer to an appropriate degree of accuracy	Your answer can only be as accurate as the values you are working with, so if the values are given to 3 significant figures you must give your answer to 3 significant figures
Give your answer in its simplest form	Commonly used when answer involves a fraction or ratio – mainly on the non-calculator paper
Simplify	Commonly used in algebra, means collect like terms
Solve	Commonly used in algebra for equations, means find the value of 'x'
Factorise	First look for common factor to put outside brackets, also used for quadratic expressions putting into 2 sets of brackets
Factorise fully	Can be a hint that there is more than 1 common factor – look for the number of marks awarded
Expand	Multiply out brackets
Expand and simplify	Multiply out brackets and then collect like terms
Measure	Use a ruler or a protractor to accurately measure lengths and/or angles
Draw an accurate diagram	Use ruler and protractor; lengths must be exact, angles must be accurate
Construct, using ruler and compasses	Use a ruler to draw any straight lines and a pair of compasses to draw arcs. Do not rub out any construction lines
Sketch	An accurate drawing is not required, can be drawn freehand
Diagram NOT accurately drawn	Commonly seen next to diagrams that are not drawn to scale, so do not measure angles or sides
Use your graph	Read values from your graph and use them
Describe fully	Usually describing transformations. You must include the following in your answer: • Translation **and** identify column vector • Reflection **and** state equation of line of symmetry • Rotation **and** identify angle with direction, (clockwise or anticlockwise) **and** identify coordinates of centre of rotation • Enlargement **and** state scale factor **and** identify coordinates of centre of enlargement
You must show all your working	Marks will be lost if working is not shown
Give a reason for your answer or explain why	Used in problem solving questions, when decisions are made they must be supported by working and a worded explanation Also used in questions involving finding angles and a written reason is required

In **geometric reasoning** questions, which involve working with properties of shapes and angles, you will be asked to give a reason for each stage of your working. To gain full marks for these questions, you must include the correct reasons for the methods you have used in your working. It is advised to include the words in **bold** in your solution when appropriate.

Lines:

Alternate angles on parallel lines are **equal**

Corresponding angles on parallel lines are **equal**

Vertically opposite angles are **equal**

Angles on a **straight line** add up to **180°**

Angles at a **point** add up to **360°**

Polygons:

Angles in a **triangle** add up to **180°**

The **exterior angle** of a **triangle** is equal to the sum of the **interior opposite angles**

Base **angles** of an **isosceles** triangle are **equal**

Angles in an **equilateral triangle** are **equal** (are 60°)

Angles in a **quadrilateral** (4-sided shape) add up to **360°**

Exterior angles of a **polygon** add up to **360°**

The **interior angle** and an **exterior angle** of a polygon add up to **180°**

Tools for success

This book will give you the information and practice that you need to do well in your GCSE Maths exam. Here are some tips and tools to maximise your success in the examination:

Before the examination

Equipment. The front page of each examination paper says you must have:

Ruler graduated in centimetres and millimetres, protractor, pair of compasses, pen, HB pencil, eraser.

In addition, on the two calculator papers you must have a calculator. It is highly likely that you will need to use all of these items across the three examination papers, so make sure you have everything prepared and ready in a clear pencil case the day before each exam.

Ruler: It is best to use a ruler that is transparent and marked with cm and mm graduations. Flexible curve rulers are **not** permitted; the same goes for French curves.

Protractor: A semi-circle that measures up to 180° version will be enough, but 360° protractors (right) are permitted. Protractors that convert degrees in a circle to percentages (e.g. 90° = 25%) are **not** permitted in the exam.

 Pair of compasses: You will need a pair of compasses for any questions to do with constructions, loci or possibly bearings. Make sure your pencil is not too thick for the pair of compasses you have.

Pen: The front of the exam paper says to use black ink or ball-point pen. Exam scripts are all scanned to be marked by examiners on computer screens and black ink shows up best; do not risk any working you have written in blue ink not being picked up by the scanners.

It is a good idea to take more than one black pen into the examination room with you, just in case something goes wrong with the first one.

Pencil: You are advised to use an HB pencil; an H pencil is too hard and might be too faint for scanners to pick up; a B pencil is too soft and may not be precise enough for graphs and diagrams. Use a sharp pencil for sketches, graphs and diagrams. Use **black** ink for everything else.

Erasers: Use to rub out pencil marks. You may need one if you make a mistake drawing sketches, graphs or diagrams. Remember, do **not** rub out construction lines!

Calculators: Most commonly available calculators can be used in GCSE examinations, including graphical calculators, but only a basic scientific calculator is necessary. If you are not sure about your calculator, ask your teacher to check for you in plenty of time whether your calculator can be used in the examination.

Make sure your calculator is fully charged before the exam. Take some spare batteries with you if you are not sure, or even a spare calculator. It is important that you are fully familiar with your calculator so that you make efficient use of it in the exam. Make sure that it is set for calculations in **degrees** for use in trigonometry questions. Remember that you are **not** allowed to take the calculator case into the exam hall.

Be aware that guidance on the front page of the calculator papers states 'If your calculator does not have a π button, take the value of π to be 3.142 unless the question instructs otherwise', but watch for questions where you are asked to give your answer in terms of π.

Tracing paper: The front of the examination paper says that tracing paper **may** be used although it does not mean you will **need** to use it. Tracing paper can be useful for questions involving transformations. Most schools will provide this in the exam hall if you ask for it. Tracing paper should not be attached to your exam paper when you hand it in at the end of the exam, so do not use it for working as it will not be seen.

Highlighter pens: You can take these into the exam if you find them useful for highlighting key words in questions. However, you must not use them to highlight your answers; they do not scan very well and can obscure the answers you are trying to highlight.

Anything else: If something is not on the list of permitted equipment, it cannot be taken into the exam hall. There are exceptions for students with special requirements who may need to take in special equipment, but this needs to be agreed with Edexcel well in advance by your school or college.

In the examination

Exam Paper: The paper has been designed so that you should have more than enough space to answer each question. Do not write outside the box around each question – that part will not be scanned and examiners will not see anything you write there.

Only ask for extra writing paper if it is absolutely necessary, for example if you have to completely re-write a question. Make sure any extra pages are securely attached to your exam paper and make a note by the question on the exam paper that you have worked out the answer elsewhere. Do not use extra writing paper for working out; all working should be written on the examination paper.

Make sure you check the last page or back page as it often has a question on it. If it says **BLANK PAGE**, that has been put there so that you know it is meant to be blank and there has not been a printing error.

You can answer the questions in any order. If you get stuck on a question and cannot continue with it because you just do not know what to do, leave it and come back to it at the end if you have time.

Formulae sheet: There will not be a formulae sheet with the examination paper. You may have been used to seeing one when trying out past papers, but from Summer 2017 you will need to know them. Make sure you have learnt the formulae you are expected to remember and be able to use well in advance of the exams. There is a list of the formulae that you need to learn off by heart at the back of this book on page 397.

Working: Make sure you write down all your working out; you'd be surprised how many marks you can lose by not doing so. Remember that examiners are looking to give you marks where they can, so make that as easy as possible for them to do. Write your calculations, working and answers clearly and logically so examiners can follow your arguments. There is no need to be embarrassed by anything you might write; all papers are marked anonymously and all papers are burned after six months.

If you need to cross anything out, just put a single line through it. Do not cross out working unless you replace it with alternative working.

Make sure answers are realistic and make sense. For example, a water bill should not be thousands of pounds, no-one runs at 400 miles per hour and buses are not 1 metre tall.

Time: You will have one and a half hours to complete all the questions on each examination paper. Ensure you make good use of all the time available – there is nothing else you can do in the exam hall except try and answer the questions on the examination paper. Always try to make a start on each question and write something down. Every minute counts and every mark you achieve might be the one which takes you from a 3 to a 4 or an 8 to a 9.

Any time you have left when you have finished tackling the questions should be used for going through your paper looking for any silly errors and to check your answers. For example, check you have copied any final line of your working into the answer space correctly, make sure simple additions and multiplications are correct. Make sure you have given a reason for your answer if it is asked for. Make sure that if, for example, you are asked 'which is the best buy on a packet of pet food?' you have clearly stated which packet is the best buy, following the results of your calculations. Most importantly, check you have actually answered the question by re-reading what you were asked to do!

If you have time, have another look at questions you were not able to answer. You could have a flash of inspiration on the next time of reading. Remember, just because you cannot answer part (a) of a question does not mean you cannot answer part (b). Every minute counts, so please do not waste them!

All this advice has been based on what past students have done, or have not done when attempting to achieve their maths GCSE qualification. If you can learn from their mistakes, it might make the difference between you achieving the grade you want and the one you do not want. Good luck!

Su Nicholson

Edexcel examiner

Thanks to Graham Cumming, Edexcel's subject advisor, for his advice and contribution.

How to use this book

We've tried to make this book as straightforward and easy to use as possible. You can work through from front to back or you can use the objectives charts and check-in questions at the start of each chapter to concentrate on the topics that you need most help with. Discuss with your teacher what will work best for you. Here are some of the features to help you succeed.

Before you start each topic, go through the list of **objectives** for the chapter and mark how confident you feel about them from not at all confident ▶ to very confident ▶▶. This will help you see how much you remember of the topic before you begin. You can go back to this list after you've worked through the chapter and see how confident you feel now.

For some chapters, there is additional 'stretch' material to try online that extends what you have learnt further. You can go to www.collins.co.uk/edexcelpost16 to downline stretch lessons where you see this marked.

Each chapter is broken down into different numbered topics. Each topic is explained clearly and simply.

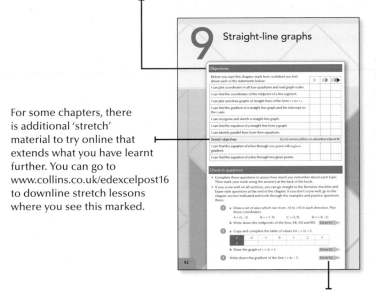

Worked **examples** are given to help you to understand the topic. You are taken through the answer step by step.

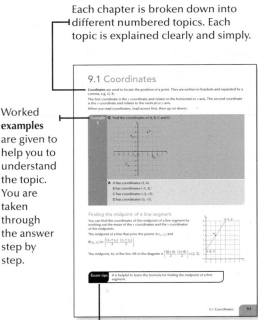

Try these **check-in questions** before you start the chapter to see how much you already know about the topic. Then mark your work using the answers at the back of the book to see which topics you did well on and which you need to brush up on. Each question relates to a section in the chapter, so look at the numbers in the arrows to see which section to go to for this refresher content.

Find out how to pick up smart marks – and not lose careless marks – on different types of questions with **exam tips** from an Edexcel examiner who knows what works.

At the end of each section there are some **questions** for you to complete to practise the skills you've learnt.

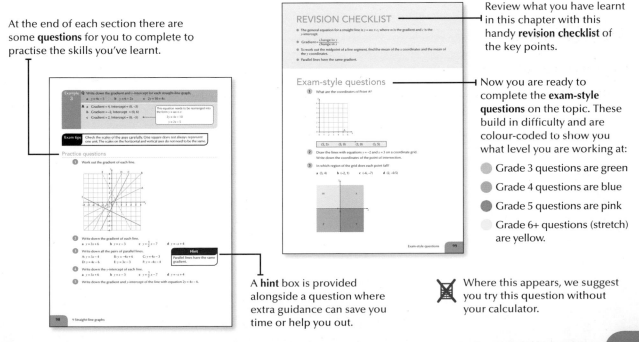

Review what you have learnt in this chapter with this handy **revision checklist** of the key points.

Now you are ready to complete the **exam-style questions** on the topic. These build in difficulty and are colour-coded to show you what level you are working at:

- Grade 3 questions are green
- Grade 4 questions are blue
- Grade 5 questions are pink
- Grade 6+ questions (stretch) are yellow.

A **hint** box is provided alongside a question where extra guidance can save you time or help you out.

Where this appears, we suggest you try this question without your calculator.

1 Numbers

Check-in questions

- Complete these questions to assess how much you remember about each topic. Then mark your work using the answers at the back of the book.
- If you score well on all sections, you can go straight to the Revision checklist and Exam-style questions at the end of the chapter. If you don't score well, go to the chapter section indicated and work though the examples and practice questions there.

1 Arrange these numbers in order of size, starting with the smallest.

  a 3603   33 060   33   36   363

  b 521   1250   2501   12 005   120

  c 64   46   −640   −406   4060   6004

  d 7340   −437   3047   −73   407

Go to 1.1 ▶

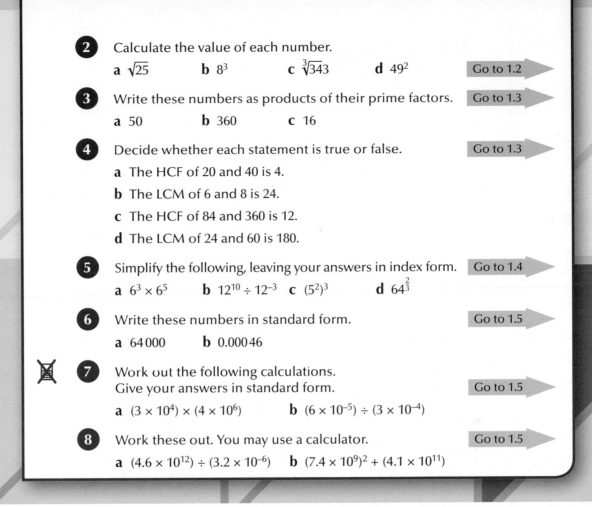

2 Calculate the value of each number.

a $\sqrt{25}$ **b** 8^3 **c** $\sqrt[3]{343}$ **d** 49^2 Go to 1.2

3 Write these numbers as products of their prime factors.   Go to 1.3

a 50   **b** 360   **c** 16

4 Decide whether each statement is true or false.   Go to 1.3

a The HCF of 20 and 40 is 4.

b The LCM of 6 and 8 is 24.

c The HCF of 84 and 360 is 12.

d The LCM of 24 and 60 is 180.

5 Simplify the following, leaving your answers in index form.   Go to 1.4

a $6^3 \times 6^5$ **b** $12^{10} \div 12^{-3}$ **c** $(5^2)^3$ **d** $64^{\frac{2}{3}}$

6 Write these numbers in standard form.   Go to 1.5

a 64 000   **b** 0.000 46

7 Work out the following calculations.
Give your answers in standard form.   Go to 1.5

a $(3 \times 10^4) \times (4 \times 10^6)$ **b** $(6 \times 10^{-5}) \div (3 \times 10^{-4})$

8 Work these out. You may use a calculator.   Go to 1.5

a $(4.6 \times 10^{12}) \div (3.2 \times 10^{-6})$ **b** $(7.4 \times 10^9)^2 + (4.1 \times 10^{11})$

1.1 Positive and negative numbers

Place value

Each digit in a number has a **place value**. The value of the digit depends on its place in the number.

The place value changes by a factor of 10 as you move from one column to the next.

Ten thousands	Thousands	Hundreds	Tens	Units
6	7	1	4	5

You read this number as sixty-seven thousand, one hundred and forty-five.

Example 1	**Q** Write these numbers in words and give the value of the underlined digit.
	a 5<u>3</u>8 **b** 23<u>7</u>1 **c** 6 3<u>5</u>2 740

A 538	five hundred and thirty-eight The digit 5 represents five hundred.
2371	two thousand, three hundred and seventy-one The digit 7 represents seven tens or seventy.
6 352 740	six million, three hundred and fifty-two thousand, seven hundred and forty The digit 5 represents fifty thousand.

Ordering integers

When ordering whole numbers:

- Put the numbers into groups with the same number of digits.
- Arrange the numbers in each group in order of size according to the place value of the digits.

Example 2	Q	Arrange these numbers in order of size. Put the smallest first. 26, 502, 794, 3297, 4209, 4351, 745 908, 5, 32, 85, 114, 54 321
	A	This becomes: 5, 26, 32, 85, 114, 502, 794, 3297, 4209, 4351, 54 321, 745 908

Example 3	Q	Gill buys some Premium Bonds for £1050. Write this figure in words.
	A	This number is one thousand and fifty pounds.

Integers are positive or negative whole numbers. Positive numbers are above zero. Negative numbers are below zero.

Integers are also **directed numbers** because they have a direction (positive or negative) as well as a size. A number without a negative sign is positive.

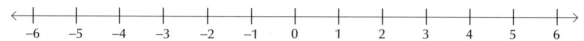

Example 4	Q	Put the correct symbol, > or <, between each pair of numbers. **a** −5 ☐ −2 **b** −4 ☐ −6 **c** 5 ☐ −2
	A	**a** −5 < −2 **b** −4 > −6 **c** 5 > −2

Directed numbers are often seen on the weather forecast in winter.

On this weather map, Aberdeen is the coldest place at −8° C and London is 6 degrees warmer than Manchester.

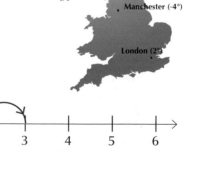

Adding and subtracting directed numbers

It can be useful to draw a number line to answer questions with negative numbers.

The temperature at 6 a.m. was −5° C.

By 10 a.m. it had risen 8 degrees.

$-5° + 8° = 3°$

So the new temperature was 3° C.

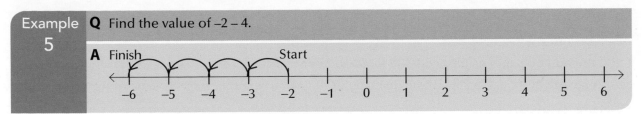

Example 5	**Q** Find the value of $-2 - 4$.
	A

When the number to be added or subtracted is negative, the normal direction of movement is reversed.

$-4 - (-3)$ is the same as $-4 + 3 = -1$

When two + signs or two − signs are together, these rules are used:

$+ (+) = +$
$- (-) = +$ } Like signs give an addition

$- (+) = -$
$+ (-) = -$ } Unlike signs give a subtraction

$-2 + (-3) = -2 - 3$
$= -5$

$-3 - (+5) = -3 - 5$
$= -8$

$6 - (-4) = 6 + 4$
$= 10$

$5 + (-2) = 5 - 2$
$= 3$

Multiplying and dividing directed numbers

Multiply and divide directed numbers as if they were positive integers and then find the sign for the answer using the following rules.

- Two like signs (both + or both −) give a positive answer.
- Two unlike signs (one + and the other −) give a negative answer.

$(+) \times (+) = +$

$(-) \times (-) = +$

$(+) \times (-) = -$

$(-) \times (+) = -$

Example 6	**Q** Work out these.
	a -6×3 **b** $-4 \times (-2)$ **c** $-20 \div (-2)$ **d** $9 \div (-3)$
	A **a** -18 **b** 8 **c** 10 **d** -3

Exam tips   You need to remember the rules of multiplying and dividing by negative numbers. You will also need to use them in algebra.

Practice questions

1 Copy and complete each statement by using < or >.
 a 2 ☐ 5   **b** 2 ☐ −5   **c** 41 ☐ 14   **d** −3 ☐ −9

2 Work out the value of these.
 a $-2 - 3$ **b** $-3 + -6$ **c** $5 - -2$ **d** $-15 + 8$

3 Which two of these have the same value?
 a $-8 + 6$ **b** $-8 \div -4$ **c** -4×-1 **d** $-6 - -4$ **e** $-28 \div -7$

4 The temperature in London at midday was 7° C. By midnight, it had fallen by 10 degrees. What was the temperature at midnight?

5 At 6 a.m. when Jeff went to work, the temperature was −3° C. By 11 a.m. it had risen to 3° C. By how many degrees had the temperature risen?

1.2 Square, cube and triangular numbers

Square numbers

Square numbers are whole numbers raised to the power of 2. For example, $5^2 = 5 \times 5 = 25$.

The first 12 square numbers are:

1	4	9	16	25	36	49	64	81	100	121	144
(1 × 1)	(2 × 2)	(3 × 3)	(4 × 4)	(5 × 5)	(6 × 6)	(7 × 7)	(8 × 8)	(9 × 9)	(10 × 10)	(11 × 11)	(12 × 12)

Square numbers can be illustrated by drawing squares:

1	4	9	16	25	...
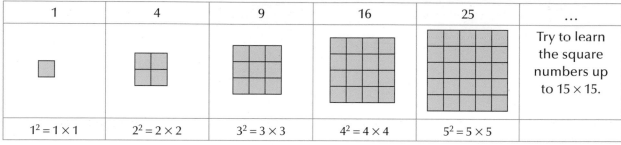					Try to learn the square numbers up to 15 × 15.
$1^2 = 1 \times 1$	$2^2 = 2 \times 2$	$3^2 = 3 \times 3$	$4^2 = 4 \times 4$	$5^2 = 5 \times 5$	

Cube numbers

Cube numbers are whole numbers raised to the power 3. For example, $5^3 = 5 \times 5 \times 5 = 125$.

1	8	27	64	125	216	...	1000
(1 × 1 × 1)	(2 × 2 × 2)	(3 × 3 × 3)	(4 × 4 × 4)	(5 × 5 × 5)	(6 × 6 × 6)	...	(10 × 10 × 10)

Cube numbers can be illustrated by drawing cubes:

1	8	27	64	125	...
$1^3 = 1 \times 1 \times 1$	$2^3 = 2 \times 2 \times 2$	$3^3 = 3 \times 3 \times 3$	$4^3 = 4 \times 4 \times 4$	$5^3 = 5 \times 5 \times 5$	

Triangular numbers

The sequence of **triangular numbers** is 1, 3, 6, 10, 15... The difference between each pair of consecutive numbers goes up by 1.

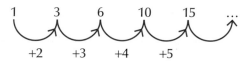

1 3 6 10 15 ...
+2 +3 +4 +5

Triangular numbers can be illustrated by drawing triangle patterns.

1	3	6	10	15	...
1	$1 + 2 = 3$	$3 + 3 = 6$	$6 + 4 = 10$	$10 + 5 = 15$	

Square roots and cube roots

$\sqrt{}$ is the square root sign. Calculating a square root is the inverse (opposite) of squaring.

Every number has two square roots, one positive and one negative.

Example 7	**Q** Find the value of $\sqrt{25}$.
	A $\sqrt{25} = 5$ or -5 since $(5)^2 = 25$ and $(-5)^2 = 25$

A **surd** is the square root of any number that is not a square number. It cannot be written exactly as a decimal.

$\sqrt{2}$, $\sqrt{3}$, $\sqrt{5}$, $\sqrt{6}$ and $\sqrt{7}$ are all surds.

$\sqrt[3]{}$ is the cube root sign. Calculating a cube root is the inverse of cubing.

Example 8	**Q** Find the values of these. **a** $\sqrt[3]{27}$ **b** $\sqrt[3]{-125}$
	A **a** $\sqrt[3]{27} = 3$ since $3 \times 3 \times 3 = 27$
	b $\sqrt[3]{-125} = -5$ since $-5 \times -5 \times -5 = -125$

Notice that when you calculate a square root you should have $\pm$ before it. When you calculate a cube root it will either be positive or negative (but not both).

Reciprocals

The **reciprocal** of a number $\frac{a}{x}$ is $\frac{x}{a}$.

Example 9	**Q** Work out the reciprocal of each of these. **a** $\frac{4}{7}$ **b** 4 **c** $\frac{x}{2}$
	A **a** $\frac{7}{4}$
	b $\frac{1}{4}$ (4 can be written as $\frac{4}{1}$)
	c $\frac{2}{x}$

To find the reciprocal of a mixed number, e.g. $1\frac{1}{2}$, first write it as an improper fraction $\left(\frac{3}{2}\right)$ and then take the reciprocal $\left(\frac{2}{3}\right)$.

See Chapter 2 for more about improper fractions.

Practice questions

1 Which of these are square numbers?

    a 48   **b** 16   **c** 1   **d** 81   **e** 400

2 Which of these are both a square number and a cube number?

    a 1   **b** 9   **c** 16   **d** 64   **e** 100

3 Work out the value of each of these.

    a $\sqrt{81}$ **b** $\sqrt[3]{64}$ **c** 1^3 **d** 100^2

4 Find the next two numbers in this sequence.

    1, 3, 6, 10, 15, ____ , ____

5 Give the reciprocal of each number.

    a 5   **b** $\frac{1}{3}$ **c** $\frac{4}{9}$ **d** $\frac{3}{20}$ **e** $5\frac{1}{5}$

1.3 Factors, multiples and primes

Factors

Factors are whole numbers that divide exactly into another number.

Example 10	**Q** Find the factors of 12.
	A The factors of 12 are 1, 2, 3, 4, 6, and 12.

Multiples

If one number is multiplied by another, the result is a **multiple** of both of the numbers. The multiples of 7, for example, are all the numbers in the 7 times table: 7, 14, 21, 28, 35...

Example 11	**Q** Find the multiples of 5.
	A Multiples of 5 are 5, 10, 15, 20, 25...
	Since: $1 \times 5 = 5$
	$2 \times 5 = 10$
	$3 \times 5 = 15$
	$4 \times 5 = 20$
	$5 \times 5 = 25$

Prime numbers

A **prime number** has exactly two factors, 1 and itself.

The prime numbers up to 20 are 2, 3, 5, 7, 11, 13, 17 and 19.

Note that 1 is not a prime number as it has only one factor and that 2 is the only even prime number.

Prime factors

Apart from prime numbers, any whole number greater than 1 can be written as a product of **prime factors**. This means the number is written using only prime numbers multiplied together.

The diagram below shows the prime factors of 60.

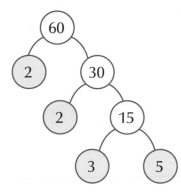

- Divide 60 by its first prime factor, 2.

- Divide 30 by its first prime factor, 2.

- Divide 15 by its first prime factor, 3.

- We can now stop because the number 5 is prime.

As a product of its prime factors, 60 is written as: $60 = 2 \times 2 \times 3 \times 5$

or in index form: $60 = 2^2 \times 3 \times 5$.

Highest common factor (HCF)

The highest factor that two numbers have in common is called the **HCF**.

Example 12	**Q** Find the HCF of 60 and 96.
	A Write the numbers as products of their prime factors. $\qquad 60 = 2 \times 2 \qquad\qquad \times 3 \times 5$ $\qquad 96 = 2 \times 2 \times 2 \times 2 \times 2 \times 3$ Ring the factors that are common. $\qquad 60 = \boxed{2} \times \boxed{2} \qquad\qquad \times \boxed{3} \times 5$ $\qquad 96 = \boxed{2} \times \boxed{2} \times 2 \times 2 \times 2 \times \boxed{3}$ The HCF of 60 and 96 is $= 2 \times 2 \times 3$ $\qquad\qquad\qquad\qquad\qquad = 12$

Lowest (least) common multiple (LCM)

The **LCM** of two (or more) numbers is the lowest number that is a multiple of the numbers.

Example 13	**Q** Find the LCM of 60 and 96.
	A Write the numbers as products of their prime factors. $\qquad 60 = 2 \times 2 \qquad\qquad \times 3 \times 5$ $\qquad 96 = 2 \times 2 \times 2 \times 2 \times 2 \times 3$ 60 and 96 have a common factor of $2 \times 2 \times 3$, so it is only counted once. $\qquad 60 = \boxed{2} \times \boxed{2} \qquad\qquad \times \boxed{3} \times 5$ $\qquad 96 = \boxed{2} \times \boxed{2} \times 2 \times 2 \times 2 \times \boxed{3}$ The LCM of 60 and 96 is $\qquad\qquad = 2 \times 2 \times 2 \times 2 \times 2 \times 3 \times 5$ $\qquad\qquad = 480$

Example 14

Q Buses to St Albans leave the bus station every 20 minutes. Buses to Hatfield leave the bus station every 14 minutes.

A bus to St Albans and a bus to Hatfield both leave the bus station at 10 a.m.

When will buses to both St Albans and Hatfield next leave the bus station at the same time?

A You need to find the LCM of 20 and 14.

$$20 = \boxed{2} \times 2 \times 5$$
$$14 = \boxed{2} \qquad \times 7$$
$$LCM = 2 \times 2 \times 5 \times 7$$
$$LCM = 140$$

Both buses will leave at the same time 140 minutes later,
i.e. 2 hours and 20 minutes later.

This is at 12:20 p.m.

You could also list both the times of the buses from Hatfield and St Albans and find the time that is the same in both lists. This is a useful check.

Buses to St Albans : 20, 40, 60, 80, 100, 120, 140 ...

Buses to Hatfield : 14, 28, 42, 56, 70, 84, 98, 112, 126, 140 ...

Both lists have 140 minutes, i.e. 2 hours and 20 minutes.

Exam tips In questions like the one above, always check that you have answered the question asked. Notice that this question not only requires you to find the '140' but then to work out 140 minutes after 10 a.m. to find the time buses will leave.

Practice questions

1 Which of these are not multiples of 3?

    a 18   **b** 25   **c** 32   **d** 304   **e** 6000

2 Which of these are prime numbers?

    a 1   **b** 2   **c** 9   **d** 17   **e** 47

3 Write each of these as a product of prime factors.

    a 24   **b** 60   **c** 96   **d** 140

4 Find the highest common factor of each pair of numbers.

    a 48 and 60   **b** 96 and 140

5 Find the lowest common multiple of each pair of numbers.

    a 24 and 140   **b** 60 and 96

6 A blue light flashes every 15 minutes. A green light flashes every 12 minutes. They both flash together at 1 p.m. When do they flash together again?

1.4 Indices

An **index** is sometimes called a **power**.

The base $\longrightarrow a^b \longleftarrow$ The index or power

Laws of indices

The laws of indices can be used for numbers and algebra. The base has to be the same when the laws of indices are applied. See Chapter 5 for applying laws of indices in algebra.

$a^n \times a^m = a^{n+m}$

$a^n \div a^m = a^{n-m}$

$(a^n)^m = a^{n \times m}$

$a^0 = 1$

$a^1 = a$

$a^{-n} = \frac{1}{a^n}$

$a^{\frac{1}{m}} = \sqrt[m]{a}$

$a^{\frac{n}{m}} = \left(\sqrt[m]{a}\right)^n$

Example 15

Q Simplify the following, leaving your answers in index form.

a $5^2 \times 5^3$ **b** $8^{-5} \times 8^{12}$ **c** $(2^3)^4$

A **a** $5^2 \times 5^3 = 5^{2+3}$
$= 5^5$

b $8^{-5} \times 8^{12} = 8^{-5+12}$
$= 8^7$

c $(2^3)^4 = 2^{3 \times 4}$
$= 2^{12}$

Example 16

Q Evaluate these. **a** 4^2 **b** 5^0 **c** 3^{-2} **d** $36^{\frac{1}{2}}$ **e** $8^{\frac{2}{3}}$ •—(Hint: Evaluate means to work out.)

A **a** $4^2 = 4 \times 4$
$= 16$

b $5^0 = 1$

c $3^{-2} = \frac{1}{3^2}$
$= \frac{1}{9}$

d $36^{\frac{1}{2}} = \sqrt{36}$
$= 6$

e $8^{\frac{2}{3}} = \left(\sqrt[3]{8}\right)^2$
$= 2^2$
$= 4$

Example 17

Q Simplify the following, leaving your answers in index form.

a $7^2 \times 7^5$ **b** $6^9 \div 6^2$ **c** $\frac{3^7 \times 3^2}{3^{10}}$ **d** $7^9 \div 7^{-10}$

A **a** 7^7 **b** 6^7 **c** $\frac{3^9}{3^{10}} = 3^{-1}$ **d** 7^{19}

Example 18

Q Evaluate these. **a** 3^3 **b** 7^0 **c** $64^{\frac{1}{3}}$ **d** $81^{\frac{1}{2}}$ **e** 5^{-2} **f** $\left(\frac{4}{9}\right)^{-2}$

A **a** $3^3 = 3 \times 3 \times 3$
$= 27$

b $7^0 = 1$

c $64^{\frac{1}{3}} = \sqrt[3]{64}$
$= 4$

d $81^{\frac{1}{2}} = \sqrt{81}$
$= 9$

e $5^{-2} = \frac{1}{5^2}$
$= \frac{1}{25}$

f $\left(\frac{4}{9}\right)^{-2} = \left(\frac{9}{4}\right)^2$
$= \frac{81}{16}$
$= 5\frac{1}{16}$

Exam tips   Questions involving negative powers are complex. Remember to make the negative power into a positive power first by taking the reciprocal.

Practice questions

1 Write each expression in index form.

a $3 \times 3 \times 3 \times 3 \times 3$ **b** $10 \times 10 \times 10$ **c** $2 \times 2 \times 2 \times 2 \times 2 \times 2$

2 Simplify each expression. **a** $5^2 \times 5^2$ **b** $7^3 \times 7^5$ **c** $2^{17} \times 2^3$ **d** $3^9 \div 3^3$

3 Rewrite each expression as a single power. **a** $6^5 \div 6^4$ **b** $8^{20} \div 8^{15}$ **c** $\frac{7^6}{7^4}$

4 Given that $3^4 = x^2$, work out the value of x.

5 Evaluate these.

a 12^0 **b** $64^{\frac{1}{2}}$ **c** $343^{\frac{1}{3}}$ **d** 8^{-1} **e** 10^{-3}

1.5 Standard index form

Standard index form (standard form) is useful for writing very large or very small numbers in a simpler way.

This is standard form for writing a number:

$$a \times 10^n$$

A number between 1 and 10 $1 \leqslant a < 10$

The value of n is the number of places the digits have to be moved to return a to the value of the original number.

If the original number is 10 or more, n is positive.

If the original number is less than 1, n is negative.

If the original number is 1 or more but less than 10, n is zero.

Example 19	**Q** Write 2 730 000 in standard form.
	A $a = 2.73$ The digits have moved 6 places between a and the original number. Since 2 730 000 > 10, n is positive. So, $2 730 000 = 2.73 \times 10^6$

Example 20	**Q** Write 0.000 046 in standard form.
	A $a = 4.6$ The digits have moved 5 places between a and the original number. Since 0.000 046 < 1, n is negative. So, $0.000 046 = 4.6 \times 10^{-5}$

Standard form on a calculator

To enter a number written in standard form into your calculator, you use:

$\boxed{\times 10^x}$ $\boxed{\text{EXP}}$ or $\boxed{\text{EE}}$

For example, $(2 \times 10^3) \times (6 \times 10^7)$ would be keyed in as:

$\boxed{2}$ $\boxed{\times 10^x}$ $\boxed{3}$ $\boxed{\times}$ $\boxed{6}$ $\boxed{\times 10^x}$ $\boxed{7}$ $\boxed{=}$

or $\boxed{2}$ $\boxed{\text{EXP}}$ $\boxed{3}$ $\boxed{\times}$ $\boxed{6}$ $\boxed{\text{EXP}}$ $\boxed{7}$ $\boxed{=}$

Check that you can use your calculator correctly to obtain the solution 1.2×10^{11}.

Calculations in standard form

🖩 Work out the following using a calculator. Check that you get the answers given here.

Example 21	**Q** **a** $(6.7 \times 10^7)^3$	**b** $\dfrac{(4 \times 10^9)}{(3 \times 10^4)^2}$	**c** $\dfrac{(5.2 \times 10^6) \times (3 \times 10^7)}{(4.2 \times 10^5)^2}$
	A **a** 3.0×10^{23} (2 s.f.)	**b** $4.\dot{4}$	**c** 884.4 (1 d.p.)

❌ On a non-calculator paper you can use laws of indices to work out the answers.

Example 22	**Q** **a** $(2 \times 10^3) \times (6 \times 10^7)$	**b** $(6 \times 10^4) \div (3 \times 10^{-2})$
	A **a** $(2 \times 10^3) \times (6 \times 10^7)$	**b** $(6 \times 10^4) \div (3 \times 10^{-2})$
	$= (2 \times 6) \times (10^3 \times 10^7)$	$= (6 \div 3) \times (10^4 \div 10^{-2})$
	$= 12 \times 10^{3+7}$	$= 2 \times 10^{4-(-2)}$
	$= 12 \times 10^{10}$	$= 2 \times 10^6$
	$= 1.2 \times 10^1 \times 10^{10}$	
	$= 1.2 \times 10^{11}$	

Example 23	**Q** $(3 \times 10^4)^2$
	A $(3 \times 10^4)^2 = (3 \times 10^4) \times (3 \times 10^4)$
	$= (3 \times 3) \times (10^4 \times 10^4)$
	$= 9 \times 10^8$

You also need to be able to work out more complex calculations.

Example 24	**Q** The mass of Saturn is 5.7×10^{26} tonnes. The mass of the Earth is 6.1×10^{21} tonnes. How many times heavier is Saturn than the Earth? Give your answer in standard form, correct to 2 significant figures.
	A $\dfrac{5.7 \times 10^{26}}{6.1 \times 10^{21}} = 93\,442.6$
	Now rewrite your answer in standard form.
	Saturn is 9.3×10^4 times heavier than the Earth (2 s.f.).

Exam tips Remember to write down the figures shown on your calculator before rounding to two significant figures. Make sure that you check that your final answer is written in standard form.

Practice questions

1 A single cold virus is $0.000\,000\,02$ m long.
Express this in standard form.

2 Work out these. **a** 1×10^4 **b** 1.2×10^5 **c** 1.23×10^{-5} **d** 1.234×10^{-1}

3 One atom of gold has a mass of $0.000\,000\,000\,000\,000\,000\,000\,000\,33$ g.
Express this in standard form.

4 Work out each of these. Give all your answers in standard form.

a $(6 \times 10^3) + (2 \times 10^2)$
b $(6 \times 10^3) - (2 \times 10^2)$
c $(6 \times 10^3) \times (2 \times 10^2)$
d $(6 \times 10^3) \div (2 \times 10^2)$

5 You have approximately 2×10^{13} red corpuscles in your bloodstream. Each red corpuscle weighs about $0.000\,000\,000\,1$ g. Work out the total mass of your red corpuscles, in kilograms. Give your answer in standard form.

REVISION CHECKLIST

- Each digit in a number has a **place value**. The value of the digit depends on its place in the number.

- Integers are whole numbers that can be positive or negative.

- When multiplying or dividing positive and negative numbers:
  - Two like signs (both + or both −) give a positive answer
  - Two unlike signs (one + and the other −) give a negative answer.

- Any non-prime whole number greater than 1 can be written as a product of its prime factors.

- The highest factor that two numbers have in common is called the highest common factor (HCF).

- The lowest number that is a multiple of two or more numbers is called the lowest (least) common multiple (LCM).

- Make sure you know and can use all the laws of indices.

- A negative power is the reciprocal of the positive power.

- Fractional indices mean roots.

- Numbers in standard form will be written as $a \times 10^n$.
  - $1 \leq a < 10$
  - n is positive when the original number is 10 or more.
  - n is negative when the original number is less than 1.
  - n is zero when the original number is 1 or more but less than 10.

Exam-style questions

1 Sam uses these cards to make three-digit numbers.
Write down a number she could make which is less than 300.

$$\boxed{8}\ \boxed{2}\ \boxed{5}$$

2 Which of these has a value which is equal to the value of $-13 + 8$?

$5 - 10$	$-11 + 6$	$-8 - -3$	$10 - 5$

3 Simplify $5 \times 5 \times 5 \times 5$. Give your answer in index form.

4 Write down the square numbers.

-4	13	120	9	1

5 Write down the value of $(-7)^{-3}$.

6 What is the reciprocal of $3\frac{1}{8}$?

7 Jess simplifies $8^4 \times 8^3$ and gives the answer as 8^{12}. Is she correct?
Give a reason for your answer.

8 Find the highest common factor of 28 and 60.

9 Write 116 as a product of prime factors. Give your answer in index form.

10 Find the lowest common multiple of 12 and 30.

11 Write down the value of $125^{\frac{1}{3}}$.

12 What is the value of 16^0?

0	1	4	−4	16

13 Which of these are surds?

$\sqrt{25}$	$5\frac{1}{8}$	$\frac{1}{3}$	$\sqrt{48}$	$\sqrt[3]{2}$

14 Which of these is not a number in standard form?

10×10^{-5}	8×10^{-6}	6.2×10^{13}	4.9×10^3	1.01×10^{14}

15 The light on Tom's smoke alarm flashes every 50 seconds. The light on Tom's house alarm flashes every two minutes. They flash together at 2 a.m. What is the next time they flash together?

16 Written as a product of prime factors, $70 = 2 \times 5 \times 7$.
Find the value of a if $700 = 2^a \times 5^a \times 7$.

17 Work out $(8 \times 10^3) + (6 \times 10^5)$. Give your answer in standard form.

18 Find the value of $(9 \times 10^5) \div (3 \times 10^3)$. Give your answer as an ordinary number.

19 The population of Iceland is 3×10^5. The population of Scotland is 5.4×10^6. How many times larger is the population of Scotland than the population of Iceland?

20 Write down the value of $\left(\frac{2}{3}\right)^{-2}$.

**Now go back to the list of objectives at the start of this chapter.
How confident do you now feel about each of them?**

2 Fractions and decimals

Check-in questions

- Complete these questions to assess how much you remember about each topic. Then mark your work using the answers at the back of the book.

- If you score well on all sections, you can go straight to the Revision checklist and Exam-style questions at the end of the chapter. If you don't score well, go to the chapter section indicated and work through the examples and practice questions there.

1 Write these fractions in order, starting with the smallest.

$\frac{3}{8}$ $\frac{7}{12}$ $\frac{1}{2}$ $\frac{4}{5}$

Go to 2.1 ▶

2 Match the pairs of equivalent fractions.

$\frac{5}{9}$ $\qquad$ $\frac{12}{36}$

$\frac{3}{10}$ $\qquad$ $\frac{24}{30}$

$\frac{4}{5}$ $\qquad$ $\frac{20}{36}$

$\frac{1}{3}$ $\qquad$ $\frac{9}{30}$

Go to 2.1 ▶

3 Work out these.

    a $\frac{2}{3} + \frac{1}{5}$ **b** $2\frac{6}{7} - \frac{1}{3}$ **c** $\frac{2}{9} \times \frac{5}{7}$ **d** $\frac{3}{11} \div \frac{22}{27}$ Go to 2.2

4 Work out these.

    a $2\frac{1}{2} + 3\frac{1}{5}$ **b** $2\frac{7}{10} - 1\frac{1}{9}$ **c** $3\frac{1}{5} \times \frac{2}{15}$ **d** $5\frac{1}{4} \div \frac{3}{8}$ Go to 2.2

5 Three-sevenths of the pages in a magazine contain advertisements. Given that 12 pages contain advertisements, work out the number of pages in the magazine.   Go to 2.3

6 Rosie watches two television programmes. The first programme lasts $\frac{3}{4}$ of an hour and the second lasts $2\frac{2}{3}$ hours. Work out the total length of the two programmes.   Go to 2.3

7 Write these fractions and decimals in order of size, starting with the smallest.

    $\frac{3}{4}$ 0.4   0.85 $\frac{8}{10}$ $\frac{675}{1000}$ Go to 2.4

2.1 Fractions

A **fraction** is part of a whole number. The top number is called the **numerator** and the bottom number the **denominator**.

In a **proper fraction**, such as $\frac{2}{5}$, the denominator is greater than the numerator.

In an **improper fraction**, such as $\frac{15}{7}$, the numerator is greater than the denominator.

A **mixed number** has a whole number part and fraction part, such as $1\frac{4}{9}$.

Equivalent fractions

Equivalent fractions have the same value. You can calculate equivalent fractions by either multiplying *or* dividing both the numerator *and* the denominator by the same number.

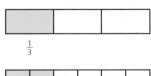

$$\overset{\times 2}{\frac{1}{3} = \frac{2}{6}}_{\times 2}$$

$$\overset{\div 2}{\frac{1}{3} = \frac{2}{6}}_{\div 2}$$

Example 1	**Q** Work out the value of x in these pairs of equivalent fractions.
	a $\frac{5}{7} = \frac{x}{35}$ **b** $\frac{40}{50} = \frac{x}{5}$
	A **a** $\overset{\times 5}{\frac{5}{7} = \frac{25}{35}}_{\times 5}, x = 25$ **b** $\overset{\div 2}{\frac{40}{50} = \frac{4}{5}}_{\div 2}, x = 4$

Simplest form

Fractions can be simplified if the numerator and denominator have a common factor.

Fractions can be written in their lowest terms by dividing the numerator and denominator by the highest common factor.

$$\frac{12}{15} = \frac{4}{5}$$

Comparing and ordering fractions

To compare or order fractions, first convert the fractions to equivalent fractions with a common denominator. The LCM of the denominators is the lowest common denominator of the fractions.

Example 2	**Q** Write the fractions $\frac{7}{8}$, $\frac{4}{5}$ and $\frac{11}{20}$ in order, starting with the smallest.
	A The LCM of 8, 5 and 20 is 40.
	$\frac{7}{8} = \frac{35}{40}$ $\frac{4}{5} = \frac{32}{40}$ $\frac{11}{20} = \frac{22}{40}$
	In order: $\frac{22}{40}$, $\frac{32}{40}$, $\frac{35}{40}$
	Reverting to their original form: $\frac{11}{20}$, $\frac{4}{5}$, $\frac{7}{8}$

Exam tips When you are asked to write fractions in order, write the fractions in your final answer in the same form as they are given in the question.

Practice questions

1 Convert each of these mixed numbers to improper fractions.

   a $3\frac{3}{10}$ **b** $2\frac{4}{9}$ **c** $5\frac{6}{11}$ **d** $13\frac{14}{23}$

2 Convert each of these improper fractions to a mixed number.

   a $\frac{48}{9}$ **b** $\frac{93}{10}$ **c** $\frac{15}{4}$ **d** $\frac{107}{2}$

3 Write each fraction in its lowest terms.

   a $\frac{8}{48}$ **b** $\frac{18}{45}$ **c** $\frac{12}{32}$ **d** $\frac{52}{208}$

4 Match each fraction on the left with its equivalent fraction in its lowest terms.

   $\frac{48}{52}$ $\frac{3}{4}$

   $\frac{13}{39}$ $\frac{1}{3}$

   $\frac{21}{28}$ $\frac{12}{13}$

5 Write these fractions in order of size, starting with the smallest.

   $\frac{7}{10}$ $\frac{3}{11}$ $\frac{3}{5}$ $\frac{9}{16}$

2.2 Calculating with fractions

Fractions of quantities

To find a fraction of a quantity, you multiply the fraction by the quantity.

Example 3	**Q** In a survey of 24 college students, $\frac{3}{8}$ prefer Reading Festival, $\frac{1}{6}$ prefer NASS and the remainder prefer Glastonbury. How many students prefer Glastonbury?
	A $\frac{3}{8} \times 24 = 9$ students prefer Reading Festival $\boxed{24 \div 8 \times 3 = 9}$
	$\frac{1}{6} \times 24 = 4$ students prefer NASS
	So $24 - (9 + 4) = 11$ students prefer Glastonbury.

> **Exam tips** To find a fraction of a quantity, divide by the denominator and multiply by the numerator.

You also need to be able to find a whole from a given fraction.

Example 4	**Q** $\frac{1}{4}$ of a length of wood is 32 cm. What is the total length of wood?
	A Let L stand for the length of wood.
	$\frac{1}{4}$ of $L = 32$ cm
	$32 \times 4 = 128$ cm $\boxed{\text{The full length is } \frac{4}{4}}$ 
	The total length of the wood is 128 cm.

Adding fractions

To add fractions with different denominators, first change them to equivalent fractions with the same denominator, then add the numerators.

Example 5	**Q** $\frac{5}{9} + \frac{1}{7} = \square$
	A The lowest common denominator is 63.
	$\frac{5}{9} + \frac{1}{7} = \frac{35}{63} + \frac{9}{63}$
	$= \frac{44}{63}$

Subtracting fractions

To subtract fractions with different denominators, first change them to equivalent fractions with the same denominator, then subtract the numerators.

Example 6

Q $\frac{4}{5} - \frac{1}{3} = \square$

A The lowest common denominator is 15.

$$\frac{4}{5} - \frac{1}{3} = \frac{12}{15} - \frac{5}{15}$$

$$= \frac{7}{15}$$

Exam tips Remember, when you add or subtract fractions with a common denominator, you only add or subtract the numerators.

Multiplying fractions

To multiply fractions, first rewrite any whole or mixed numbers as improper (top-heavy) fractions if necessary. Next multiply together the numerators, then the denominators.

Example 7

Q $\frac{2}{7} \times \frac{4}{5} = \square$

A $\frac{2}{7} \times \frac{4}{5} = \frac{2 \times 4}{7 \times 5}$

$$= \frac{8}{35}$$

Dividing fractions

To divide fractions, first rewrite any whole or mixed numbers as improper fractions.

Then multiply the first fraction by the reciprocal of the second fraction.

Example 8

Q $2\frac{1}{3} \div 1\frac{2}{7} = \square$

A $2\frac{1}{3} \div 1\frac{2}{7} = \frac{7}{3} \div \frac{9}{7}$

$$= \frac{7}{3} \times \frac{7}{9}$$

$$= \frac{49}{27}$$

$$= 1\frac{22}{27} \bullet \longrightarrow \boxed{\text{Rewrite the answer as a mixed number.}}$$

Practice questions

1 How much more is $\frac{3}{4}$ of £120 than $\frac{2}{5}$ of £80?

2 Work out these additions.

    a $\frac{2}{5} + \frac{3}{8}$ **b** $\frac{4}{7} + \frac{1}{4}$ **c** $1\frac{9}{10} + \frac{7}{8}$

3 Work out these subtractions.

    a $\frac{7}{12} - \frac{1}{3}$ **b** $\frac{10}{11} - \frac{5}{6}$ **c** $1\frac{1}{2} - \frac{7}{10}$

2 Fractions and decimals

4 Work out these.

   a $\dfrac{5}{9} \times \dfrac{3}{4}$ **b** $\dfrac{2}{3} \times \dfrac{3}{4}$ **c** $4\dfrac{1}{2} \times 1\dfrac{7}{9}$

5 Complete these division calculations.

   a $\dfrac{7}{8} \div \dfrac{5}{16}$ **b** $\dfrac{6}{7} \div 3$ **c** $5\dfrac{1}{6} \div 4\dfrac{4}{9}$

2.3 Fraction problems

You may also be asked to solve problems involving fractions.

Example 9

Q A school has 1400 students. 740 students are boys.

$\frac{3}{5}$ of the boys and $\frac{1}{4}$ of the girls study French.

Work out the total number of students in the school that study French.

A $\dfrac{3}{5} \times 740 = 444$ boys study French

1400 − 740 = 660 girls in the school

$\dfrac{1}{4} \times 660 = 165$ girls study French

444 + 165 = 609 students study French

Example 10

Q Charlotte's take-home pay is £930. She gives her mother $\frac{1}{3}$ of this and spends a further $\frac{1}{5}$ on going out. What fraction of the £930 is left?

Give your answer as a fraction in its simplest form.

A First work out what fraction she has spent.

$$\dfrac{1}{3} + \dfrac{1}{5} = \dfrac{5}{15} + \dfrac{3}{15}$$

$$= \dfrac{8}{15}$$

Then work out what fraction of the whole is left.

$$1 - \dfrac{8}{15} = \dfrac{7}{15}$$ ●————(The fraction is already its simplest form.)

$\dfrac{7}{15}$ of the £930 is left.

Exam tips Always re-read the question to check that you have answered it correctly.

Practice questions

1 Sarra plans to spend $\frac{1}{3}$ of her wages on rent, $\frac{2}{5}$ on bills (food, electricity, etc.), $\frac{1}{6}$ on travel and $\frac{1}{5}$ on going out. Is this possible? Give a reason for your answer.

2 Brin's garden measures $\frac{9}{10}$ m by $2\frac{1}{2}$ m. He has enough grass seed to cover 2 m². Will this cover the whole garden? Show your working.

3 Charlie earns £800 per month. She pays £350 rent and £150 for bills each month. What fraction of her monthly earnings is left?

4 Judy works in a hair salon at the weekends. She works from 8 a.m. to 6 p.m. on Saturday and from 10 a.m. to 3 p.m. on Sunday. Before noon on each day, she cleans the salon and chats with the customers. What fraction of her working time is this?

5 Toby has 24 days holiday per year. He plans to use $\frac{1}{2}$ of those in the summer and $\frac{1}{3}$ of them at Christmas. How many days holiday will he have left?

2.4 Decimals and fractions

When a number is written in decimal form, a decimal point separates the whole number part from the part that represents the fraction.

Thousands	Hundreds	Tens	Unit		Tenths	Hundredths	Thousandths
6	7	1	4	.	2	3	8

- To change a fraction into a decimal, divide the numerator by the denominator. You can do this using a calculator or by short division.

For example: $\frac{2}{5} = 2 \div 5 = 0.4$ $\frac{1}{8} = 1 \div 8 = 0.125$

The examples above are **terminating** decimals. Some other fractions give **recurring** decimals. These never stop, but sometimes have a repeating pattern.

For example: $\frac{1}{3} = 0.333\,333\ldots$ usually written as $0.\dot{3}$

$\frac{5}{11} = 0.454\,545\ldots$ usually written as $0.\dot{4}\dot{5}$

$\frac{4}{7} = 0.571\,428\,571\ldots$ usually written as $0.\dot{5}7142\dot{8}$

- To change a terminating decimal into a fraction, look at the number of digits after the decimal point. The digits after the decimal point will be the numerator. The denominator will be 10 if there is one decimal place, 100 if there are two decimal places, 1000 if there are three decimal places … Then cancel, if possible.

For example: $0.23 = \frac{23}{100}$

$0.165 = \frac{165}{1000}$

$= \frac{33}{200}$

Decimal scales

Decimals are often used when reading scales. Measuring jugs, rulers and weighing scales are examples of scales that use decimals.

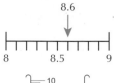

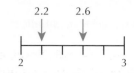

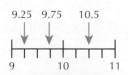

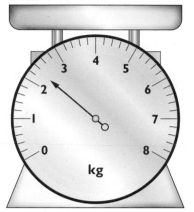

There are 10 spaces between the 8 and the 9. Each space is 0.1.

There are five spaces between the 2 and the 3. Each space is 0.2.

There are four spaces between the 9 and the 10. Each space is 0.25.

Ordering decimals

To order decimals:

- Write the numbers in columns and add any zero place fillers so that all the numbers have the same number of digits after the decimal point.

- Compare the digits in the first place of difference to decide which number is larger.

Example 11	**Q** Arrange these numbers in order of size, smallest first.
	5.29, 5.041, 5.7, 2.93, 5.71

A First write them in columns and add zero place fillers as required.

5.	2	9	0
5.	0	4	1
5.	7	0	0
2.	9	3	0
5.	7	1	0

Then reorder them: 2.930, 5.041, 5.290, 5.700, 5.710

Rewrite in original form: 2.93, 5.041, 5.29, 5.7, 5.71

Practice questions

1 Write each fraction as a decimal.

a $\frac{4}{5}$ b $\frac{9}{20}$ c $\frac{3}{16}$ d $\frac{27}{50}$

2 Write each decimal as a fraction.

a 0.8   b 0.72   c 1.35   d 2.876

3 Write down the numbers that are, or give, recurring decimals.

0.33 $\frac{1}{6}$ $\frac{3}{5}$ $\frac{4}{7}$ π

4 Write these numbers in order of size. Start with the smallest.

    2.41   4.21   1.24   1.024   2.014

5 Write these numbers in numerical order. Start with the largest.

    2.401   2.104   2.14   2.041   2.014

REVISION CHECKLIST

- In a **proper fraction**, the denominator is larger than the numerator.
- In an **improper fraction**, the numerator is larger than the denominator.
- Numbers with a whole number part and a fraction part, such as $1\frac{5}{7}$, are called mixed **numbers**.
- To add or subtract fractions, first rewrite them with a common denominator if necessary, then add or subtract only the numerators.
- When multiplying or dividing fractions, first rewrite any whole or mixed numbers as improper fractions.
    - To multiply fractions, multiply the numerators and multiply the denominators.
    - To divide fractions, multiply the first fraction by the reciprocal of the second fraction.
- To change a fraction into a decimal, divide the numerator by the denominator using short division or a calculator.
- Decimals that never stop are **recurring** decimals.
- To change a terminating decimal into a fraction, write the digits after the decimal point as the numerator over a denominator of 10 (one decimal place), 100 (two decimal places), 1000 (three decimal places)…, then cancel if necessary.
- When ordering decimals:
    - Write them in columns with the same number of decimal digits.
    - Compare the first digits of difference.

Exam-style questions

1 What fraction of an hour is 10 minutes?

2 Which decimal has the same value as $\frac{9}{20}$?

    9.20   20.9   0.45   0.920   0.05

3 Write 0.36 as a fraction in its lowest terms.

4 A book has 400 pages. 75 of these pages have colour pictures. What fraction of the pages in the book have colour pictures?

5 Jorges and Tim both went to a job interview. They were each given a test to complete. Jorges scored $\frac{17}{20}$ and Tim scored $\frac{13}{16}$. Which score is better? Show your working.

6 Tessi is the manager of a band. She is having 100 t-shirts printed. She wants $\frac{2}{5}$ of them to have only the band name, half to have only the band logo and the remainder to have both the logo and the name. How many t-shirts will have both the logo and the name?

7 What is $\frac{7}{9} - \frac{4}{5}$?

8 Calculate: $\frac{2}{9} + 5\frac{1}{4}$

9 Sue needs hair ribbons for the girls in her dance show. Each ribbon needs to be $\frac{3}{20}$ m long. She has a reel of ribbon 5 m in length. How many ribbons can she cut from the reel?

10 Write these fractions in order of size. Start with the smallest.

$\frac{5}{7}$ $\frac{7}{10}$ $\frac{29}{35}$ $\frac{4}{5}$

11 Jeff's running stride is $1\frac{1}{10}$ m. If he runs 8 km, how many strides will he have taken?

12 What is $3\frac{3}{8} \times 1\frac{4}{9}$?

13 Write these numbers in order. Start with the largest.

12.021 1.21 11.2 11.02 11.102

14 Which of these fractions is larger? Show your working.

$\frac{7}{10}$ $\frac{29}{40}$

15 What is $\frac{7}{8} + 1\frac{2}{5} \times 2\frac{3}{4}$?

16 Write these in numerical order, starting with the smallest.

0.2 $\frac{3}{20}$ 0.13 $\frac{1}{4}$

17 Which of these numbers are or will be recurring decimals.

$\frac{5}{6}$ 0.33 $\frac{1}{7}$ $\frac{9}{20}$

18 A travel company surveys fans at a concert about how they had travelled to the venue. $\frac{2}{3}$ travelled by car, $\frac{1}{5}$ travelled by train or bus and the remainder did not give an answer. Work out the fraction of the fans that did not answer.

19 Miguel runs an ice cream shop near the sea. Last week he used the amount of ice cream shown in the table.

Flavour	Number of tubs
Vanilla	12
Chocolate	15
Cherry	2
Mint	7
Strawberry	4

He is placing an order for next week. What fraction of his order should be for chocolate ice cream? Give your answer in its lowest terms.

20 A pub seating area measures $3\frac{3}{4}$ m by $4\frac{2}{5}$ m. Des needs to stain the seating area. Each tub of wood stain will cover 3 m² and costs £8.75.

How much will it cost to buy enough tubs to stain the seating area?

Now go back to the list of objectives at the start of this chapter.

How confident do you now feel about each of them?

3 Calculations

Check-in questions

- Complete these questions to assess how much you remember about each topic. Then mark your work using the answers at the back of the book.
- If you score well on all sections, you can go straight to the Revision checklist and Exam-style questions at the end of the chapter. If you don't score well, go to the chapter section indicated and work through the examples and practice questions there.

1. Copy and complete the table.

Start number	× 10	× 100	× 1000
75			
184			
3			
0.6			
0.07			
1.758			

Go to 3.1 ▶

2 Copy and complete the table.

Start number	÷ 10	÷ 100	÷ 1000
75			
184			
3			
0.6			
0.07			
1.758			

Go to 3.1

3 Work out each calculation: Go to 3.2

    a $486.5 + 92.61$ **b** $784 - 136.49$ **c** 92.6×27 **d** $2210 \div 6.8$

4 Work out: Go to 3.3

    a $48 + 12 \div 6$ **b** $26 - 5^2$ **c** $(7-3)^2 \div 4 + 8$ **d** $49 - 7 \times 4$

5 Use your calculator to work out: Go to 3.4

    a $5 + 4 \times 8.3$ **b** $5^2 + \dfrac{38}{1.8}$ **c** $\dfrac{11.1 \times 12.4}{5.6 + 3.2}$ **d** $\dfrac{3^5 - 2^6}{42}$

3.1 Multiplying and dividing numbers by power of 10

To multiply by 10, 100, 1000…, move the digits one, two, three… places to the left and write in zero place holders if necessary.

Example 1

Q Work out:

    a 5.6×10 **b** 45×10 **c** 19.3×100 **d** 78×1000

A a

H	T	U	Tenths
		5	6
	5	6	

Move the digits one place to the left.

b

H	T	U
	4	5
4	5	0

Move the digits one place to the left and add a zero in the units place.

c

Th	H	T	U	Tenths
		1	9	. 3
1	9	3	0	

Move the digits two places to the left and add a zero in the units place.

d

Tth	Th	H	T	U
			7	8
7	8	0	0	0

Move the digits three places to the left and add zeros in the units, tens and hundreds positions.

You can use the same rules when you need to multiply by multiples of 10.

Example 2

Q Work out:

a 50×30 **b** 5.2×20

A **a** $50 \times 30 = 50 \times 3 \times 10$
      $= 150 \times 10$
      $= 1500$

    b $5.2 \times 20 = 5.2 \times 2 \times 10$
      $= 10.4 \times 10$
      $= 104$

To divide by 10, 100, 1000…, move the digits one, two, three… places to the right and write in any zero place fillers if necessary.

Example 3

Q Work out:

a $17.8 \div 10$ **b** $84 \div 100$ **c** $19.6 \div 1000$

A **a** $17.8 \div 10 = 1.78$ **b** $84 \div 100 = 0.84$ **c** $19.6 \div 1000 = 0.0196$

You can use the same rules when you need to divide by multiples of 10.

Example 4

Q Work out:

a $8000 \div 20$ **b** $9.3 \div 30$

A **a** $8000 \div 20 = 8000 \div 2 \div 10$
      $= 4000 \div 10$
      $= 400$

    b $9.3 \div 30 = 9.3 \div 3 \div 10$
      $= 3.1 \div 10$
      $= 0.31$

When you multiply by a number between 0 and 1, the answer is smaller than the starting value.

Example 5

Q Work out:

a 7×0.1 **b** 7×0.01 **c** 7×0.001

A **a** $7 \times 0.1 = 0.7$ **b** $7 \times 0.01 = 0.07$ **c** $7 \times 0.001 = 0.007$

When you divide by a number between 0 and 1, the answer is bigger than the starting value.

Example 6	Q Work out:		
	a 9 ÷ 0.1	**b** 9 ÷ 0.01	**c** 9 ÷ 0.001
	A **a** 9 ÷ 0.1 = 90	**b** 9 ÷ 0.01 = 900	**c** 9 ÷ 0.001 = 9000

Practice questions

1 Multiply each number by: **i** 10   **ii** 1000

    a 63   **b** 4.21   **c** 392   **d** 0.065

2 Divide each number by: **i** 100   **ii** 1000

    a 215   **b** 63   **c** 0.97   **d** 8542

3 Match each calculation with the correct answer.

    9.3 × 0.001   0.093

    0.93 ÷ 10   0.0093

    93 × 100   9.3

    93 ÷ 10   9300

4 Which of these calculations does not have the same value as the others?

    3.8 × 10   0.38 ÷ 0.01   380 ÷ 100   0.038 × 1000

5 Lucy is helping at a maths club. The question says to work out the value of 54 ÷ 0.001.

    Lucy needs to explain whether the answer will be bigger or smaller than the starting value, 54. Write an explanation for her. Remember to give examples.

3.2 Written methods

Addition

When adding integers and decimals, the place values must line up in columns.

The same method can be used to add decimal numbers.

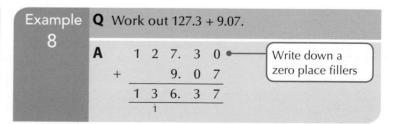

Example 7	Q Work out 5279 + 408.
	A 5 2 7 9
	+ 4 0 8
	5 6 8 7
	1

Example 8	Q Work out 127.3 + 9.07.
	A 1 2 7. 3 0 [Write down a zero place fillers]
	+ 9. 0 7
	1 3 6. 3 7
	1

Exam tips   Make sure the decimal points are in line.

Subtraction

When subtracting integers and decimals, the place values must line up in columns. Subtracting is also known as finding the **difference**.

Example 9

Q Work out 2791 − 365.

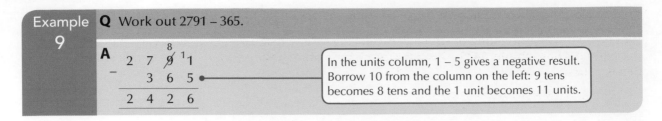

A

$$
\begin{array}{r}
2\ 7\ \overset{8}{\cancel{9}}\ ^{1}1 \\
-\quad 3\ 6\ 5 \\
\hline
2\ 4\ 2\ 6
\end{array}
$$

In the units column, 1 − 5 gives a negative result. Borrow 10 from the column on the left: 9 tens becomes 8 tens and the 1 unit becomes 11 units.

Multiplication

Multiplication is much quicker if you know your multiplication tables. You need to know how to multiply using a formal method, but grid methods are also helpful.

Example 10

Q What is 6.24 × 8?

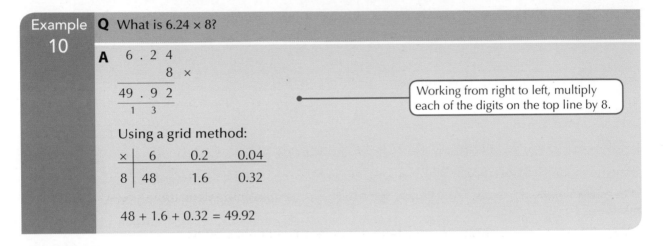

A

$$
\begin{array}{r}
6\ .\ 2\ 4 \\
8\ \times \\
\hline
49\ .\ 9\ 2 \\
{}_{1}\quad {}_{3}
\end{array}
$$

Working from right to left, multiply each of the digits on the top line by 8.

Using a grid method:

×	6	0.2	0.04
8	48	1.6	0.32

48 + 1.6 + 0.32 = 49.92

Exam tips Check your answer by estimation, rounding to 1 significant figure:
6.24 × 8 ≈ 6 × 8 = 48

When you multiply together two or more numbers, you are finding the **product** of those numbers.

Example 11

Q Work out the product of 1.89 and 23.

A Use long multiplication to solve this. Since it is more straightforward to multiply whole numbers, you can first multiply 1.89 by 100 to eliminate the decimal point.

$$
\begin{array}{r}
1\ 8\ 9 \\
\times\quad 2\ 3 \\
\hline
5\ 6\ 7 \\
3\ 7\ 8\ 0 \\
\hline
4\ 3\ 4\ 7
\end{array}
$$

189 × 3
189 × 20

Since you multiplied the number in the question by 100, you will need to divide the answer by 100.

4347 ÷ 100 = 43.47

Exam tips You can check the **size** of your answer using estimation:
1.89 × 23 ≈ 2 × 20 = 40

Division

When you divide using long or short division, take care to make sure that important zeros are not missed out. For example, $1435 \div 7 = 205$, not 25. Estimating before you divide will help you to avoid these errors.

Chunking can also be used to divide.

Example 12	Q	A bar of chocolate costs 74p. Carys has £9.82.

a What is the maximum number of bars Carys can buy?

b How much money does she have left?

A **a**

$$
\begin{array}{r}
1\ 3 \\
74\ \overline{)\ 9\ 8\ 2} \\
7\ 4 \\
\hline
2\ 4\ 2 \\
-\ 2\ 2\ 2 \\
\hline
2\ 0
\end{array}
$$

$98 \div 74 = 1$ remainder 24
now bring down the 2

$242 \div 74 = 3$ remainder 20

Carys can buy 13 bars.

b She has 20p left over.

In the example above, 74 is the **divisor**, 13 is the **quotient** and 20p is the **remainder**.

If the question had been $982 \div 74$, the answer could be written as $13\frac{20}{74}$ or $13\frac{10}{37}$.

When dividing by decimals, it is often simpler to write an equivalent calculation using whole numbers. For example:

$372.8 \div 0.4$ is the same as $3728 \div 4$ since both numbers are multiplied by 10.

$527.1 \div 0.02$ is the same as $52\,710 \div 2$ since both numbers are multiplied by 100.

Calculating the answers to some word problems may involve several steps or calculations.

Example 13	Q	This is a menu in a coffee shop.

Latte	£2.45
Cappuccino	£2.20
Frappuccino	£2.60
Panini	£3.70
Cheese and ham toastie	£3.45

Harry and his friends buy 2 lattes, 1 cappuccino, 3 paninis and 1 cheese and ham toastie. Harry hands over £30. How much change should he get?

A Start by calculating the total cost.

$2 \times £2.45 = £4.90$

$3 \times £3.70 = £11.10$

$$
\begin{array}{r}
1\ 1.\ 1\ 0 \\
4.\ 9\ 0 \\
3.\ 4\ 5 \\
+\quad 2.\ 2\ 0 \\
\hline
2\ 1.\ 6\ 5 \\
\hline
1\quad 1
\end{array}
$$

Then work out the change received: $£30.00 - £21.65 = £8.35$

Practice questions

1 Work out:

   a 5069 + 423   **b** 31.46 + 123.4   **c** 56.28 − 0.74   **d** 63 − 0.567

2 Work out:

   a 59.7 × 84   **b** 126 × 3.58   **c** 512 ÷ 40   **d** 250 ÷ 0.05

3 Libby is holding a sewing class for 24 people. Each person will need one zip. Each zip costs 34p. How much will Libby pay altogether for the zips?

4 Lou is a registrar for births. Parents can choose to pay £4.50 for a commemorative certificate. At the end of one day, Lou has been paid £58.50 for certificates. How many certificates were bought?

5 Malcom takes orders for lunch at work. People can choose from the following items.

Sandwiches	£	Drinks	£
Cheese	2.75	Cola	0.70
Ham	3.00	Lemonade	0.70
Tuna	3.20	Latte	1.80
Prawn	3.50	Hot chocolate	1.90
BLT	3.50	Tea	1.30

These are the orders on Tuesday.

Sandwiches	Number	Drinks	Number
Cheese	5	Cola	4
Ham	3	Lemonade	2
Tuna	6	Latte	3
Prawn	1	Hot chocolate	7
BLT	2	Tea	1

Malcom pays for lunch with four £20 notes. How much change should he receive?

3.3 Order of operations

BIDMAS is a mnemonic to help you remember the order in which you should complete calculations.

B	I	D	M	A	S
Brackets	Indices (powers)	Division	Multiplication	Addition	Subtraction

This means that you work out calculations in brackets first, followed by indices and, in the absence of brackets, division and multiplication calculations before addition and subtraction.

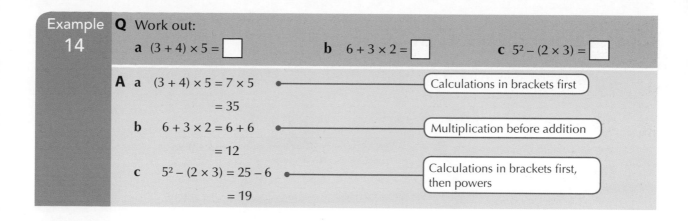

Example 14

Q Work out:

a $(3 + 4) \times 5 = \boxed{}$ **b** $6 + 3 \times 2 = \boxed{}$ **c** $5^2 - (2 \times 3) = \boxed{}$

A **a** $(3 + 4) \times 5 = 7 \times 5$ •————— Calculations in brackets first

             $= 35$

b $6 + 3 \times 2 = 6 + 6$ •————— Multiplication before addition

             $= 12$

c $5^2 - (2 \times 3) = 25 - 6$ •————— Calculations in brackets first, then powers

             $= 19$

Exam tips   The order of operations must be used for numerical and algebraic calculations.

Practice questions

1   Work out:

   a $5 + 3 \times 2$ **b** $16 \div (5 + 3) - 2$ **c** $4 \times 2 + 7$ **d** $24 - 12 \div 3$

2   Work out:

   a $2 + 100 \div 4$ **b** $12 + 3 \div 3 + 2$ **c** $16 - (4 \times 3) + 6$ **d** $100 - 50 \div 5^2$

3   Put brackets in these calculations where necessary to make each answer correct.

   a $5 \times 6 + 2 = 32$ **b** $16 - 6 \div 2 = 13$ **c** $3 \times 15 - 1 = 42$

4   Lior has to answer this question on a test.

> Work out the value of $8 + 3.2^2 - 16 \div 4$.

   His answer 0.56

   Is he correct? Why or why not?

5   Match each calculation with the correct answer.

   $6 + 4 \times 8$ 37

   $52 - 3 \times 5 + 18$ 55

   $5^2 + 4 \times 2$ 38

   $5 + 4^2 \times 2$ 33

3.4 Using a calculator

Make sure you know how to use your calculator. The diagram shows some important keys.

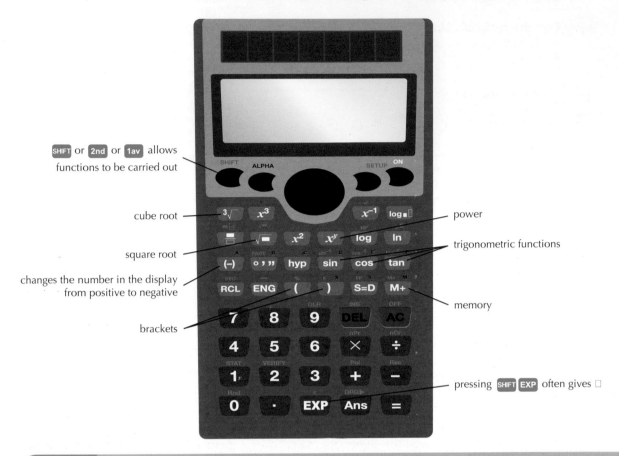

SHIFT or 2nd or 1av allows functions to be carried out

cube root

square root

changes the number in the display from positive to negative

brackets

power

trigonometric functions

memory

pressing SHIFT EXP often gives ▯

Example 15	**Q** Use your calculator to work out $\dfrac{15 \times 10 + 46}{9.2 \times 2.1}$.

A You can key this in:

(1 5 × 1 0 + 4 6) ▭ (9 . 2 × 2 . 1) =

The answer is 10.1449… Check this using your calculator.

You can also use the memory keys for the calculation above. Try writing down the key sequence for yourself.

Calculating powers and reciprocals

y^x, x^y or $x^{\wedge 2}$ are used to calculate powers, such as 2^7.

Example 16	**Q** Use the power key on the calculator to work out the value of 2^7.

A 2 x^y 7 =

The answer is 128. Check this using your calculator.

x^{-1} or $1/x$ is the **reciprocal** key. It is used to calculate the reciprocal of a number.

11 Find the value of $8 + 6^2 - 12$.

12 Find the value of $12 \div 2 + 4$.

13 Copy and complete the statement by writing $\div$, $\times$, $-$ or $+$ in the boxes.

$$5 \; \boxed{} \; 3 \; \boxed{} \; 2 \; = \; 4$$

14 Is this statement true or false?

$$\boxed{6 \times 5 - 2 = 18}$$

You must explain your answer.

15 Copy and complete the table, using a tick to show whether each statement is true or false.

	True	False
$3 - 4 \times 2 = -2$		
$5 + 6 \times 7 = 77$		
$3 + 2 \times 4 - 5 = 6$		
$25 \div 5^2 = 1$		

16 What is the reciprocal of 25? Give your answer as a decimal.

17 Which has the largest value?

$$\boxed{\text{Reciprocal of 4} \qquad 0.2 \qquad 2.4 \times 0.1 \qquad \frac{1}{4.8}}$$

18 Copy and complete the statement, writing < or > in place of the box.

$$\frac{1}{52} \; \boxed{} \; 2 \times 0.01$$

19 Suki is trying to find the reciprocal of $\frac{5}{9}$. She says it will be smaller than $\frac{5}{9}$ because a reciprocal of a number is always smaller than the number itself. Is she correct? Give a reason for your answer.

20 Write these in order of size. Start with the smallest.

$$\boxed{2^7 \qquad 3^{3.4} \qquad 4^3 \qquad 5^{2.3}}$$

Now go back to the list of objectives at the start of this chapter.

How confident do you now feel about each of them?

4 Accuracy

Objectives

Before you start this chapter, mark how confident you feel about each of the statements below:

	▶	▶▶	▶▶▶
I can round numbers and measures to the nearest 10, 100 and 1000.			
I can round numbers and measures to a specified number of decimal places.			
I can round numbers and measures to a specified number of significant figures.			
I can use inequality notation to specify simple error intervals due to truncation or rounding.			
I can apply and interpret limits of accuracy.			
I can estimate the answers to calculations.			

Check-in questions

- Complete these questions to assess how much you remember about each topic. Then mark your work using the answers at the back of the book.
- If you score well on all sections, you can go straight to the Revision checklist and Exam-style questions at the end of the chapter. If you don't score well, go to the chapter section indicated and work through the examples and practice questions there.

1 What is 3724 rounded to 2 significant figures? Choose from the options below.

| 3800 | 37 | 38 | 3700 |

Go to 4.1

2 Decide whether each statement is true or false.

Go to 4.1

  a 4625 rounded to 3 s.f. is 4630

  b 2.795 rounded to 1 d.p. is 2.7

  c 0.00527 rounded to 2 s.f. is 0.0053

  d 37 062 has 4 significant figures

3 Wallace's mass is 70 kg, correct to the nearest kilogram. Which of these is the lower bound for his mass?

| 65 kg | 65.5 kg | 69 kg | 69.5 kg |
| 70.5 kg | 74.9 kg | | |

Go to 4.1

4 Estimate the value of $\dfrac{426+37}{5.2^2-4.56}$.

Go to 4.2

4.1 Rounding

Rounding to the nearest 1, 10, 100 and 1000

When rounding to the nearest whole number, look at the digit to the right of the decimal point. If the digit is 5 or more, round up to the next whole number. If the digit is 4 or less, the whole number stays the same and the value rounds down.

To round to the nearest 10, look at the digit in the units column.

To round to the nearest 100, look at the value of the tens digit.

To round to the nearest 1000, look at the value of the hundreds digit. In each case, if the digit you look at is 5 or more, round up to the next 10, 100 or 1000. If the digit is 4 or less, the value of the number rounds down.

2469.1 is
- 2469 to the nearest whole number
- 2470 to the nearest 10
- 2500 to the nearest 100
- 2000 to the nearest 1000

Exam tips Remember to use zeros as place fillers.

Rounding to decimal places

When rounding numbers to a given number of **decimal places** (d.p.), count the given number of places to the right of the decimal point, then look at the next digit on the right. As with whole numbers, if the digit is 5 or more, round up. If the digit is 4 or smaller, the digit at the given place value stays the same and the number rounds down.

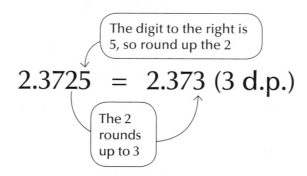

The digit to the right is 5, so round up the 2

$$2.3725 = 2.373 \text{ (3 d.p.)}$$

The 2 rounds up to 3

Example 1	**Q** Round each number as indicated in the brackets.

Q Round each number as indicated in the brackets.

a 4.6931 (2 d.p).	**b** 27.325 (2 d.p.)
c 149.3867 (3 d.p.)	**d** 271.74 (1 d.p.)

A
- **a** 4.69 — 3 < 5, so the 9 does not change
- **b** 27.33 — 5 ≥ 5, so 2 rounds up to 3
- **c** 149.387 — 7 ≥ 5, so 6 rounds up to 7
- **d** 271.7 — 4 < 5, so the 7 does not change

Rounding to significant figures

The first **significant figure** (s.f.) is the first non-zero digit from the left. The second, third … significant figures follow immediatetly after the first significant figure. They may or may not be zeros.

You use the same rules for rounding to significant figures as you do for rounding to decimal places. However, with significant figures you must replace rounded digits with zero place fillers.

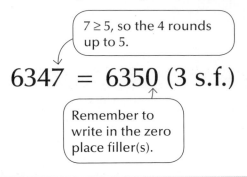

$7 \geq 5$, so the 4 rounds up to 5.

$$6347 = 6350 \ (3 \text{ s.f.})$$

Remember to write in the zero place filler(s).

Example 2

Q Round 9.3156 to

   i 3 s.f.   **ii** 2 s.f.   **iii** 1 s.f.

A i 9.32   **ii** 9.3   **iii** 9

Example 3

Q Round 0.735 to   **i** 2 s.f.   **ii** 1 s.f.

A 7 is the first significant figure. The zero before the decimal point is not significant, but must be included to preserve the place value.

   i 0.74   **ii** 0.7

Possible error of half a unit when rounding

The **upper bound** is the upper limit for a measurement.

The **lower bound** is the minimum possible value for a measurement.

If a measurement is accurate to some given amount, then the actual value lies within half a unit of that amount. So, if the mass (m) of a cat is 8.3 kg to the nearest tenth of a kilogram, then the actual mass is between 8.25 kg and 8.35 kg.

The limits of accuracy can be written using inequalities:

The error interval is the range of values between the upper and lower bounds, in which the precise value could be:

Error interval

lower bound $\geq m \leq$ upper bound

$8.25 \leq m < 8.35$

8.2   8.3   8.4

lower bound   upper bound

Example 4

Q The mass of a book is 28 g to the nearest gram. Write down the lower bound of the mass of the book.

A Let **m** represent the mass of the book.

$27.5 \leq m < 28.5$

The lower bound of the mass of the book is 27.5 g.

Truncation

Truncation is a way to approximate a number without using the rules of rounding. You round down at the given unit.

Example 1

451.73 truncated to the nearest whole number is 451.

3123.41 truncated to 1 decimal place is 3123.4

453.1678 truncated to 3 decimal places is 453.167

Example	**Q** A distance, d, is given as 4.73m truncated to 2 decimal places. Write down the error interval for the distance, d.
5	
	A $4.73 \leq m < 4.74$

Practice questions

1 Copy and complete the table.

	Rounded to 1 d.p.	Rounded to 2 d.p.	Rounded to 3 d.p.
27.1647			
893.4078			
0.0071			
−3.0414			

2 Copy and complete the table.

	Rounded to 1 s.f.	Rounded to 2 s.f.	Rounded to 4 s.f.
4096			
12 304			
679			
134			

3 Sam organises the village fair. She collects money from different stalls throughout the day and keeps an approximate total of how much money has come in. She rounds each amount to the nearest pound.

Round each amount to the nearest pound and calculate her estimated total.

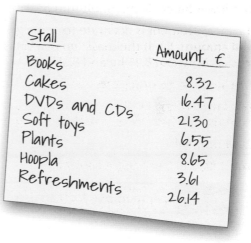

Stall — Amount, £
Books
Cakes — 8.32
DVDs and CDs — 16.47
Soft toys — 21.30
Plants — 6.55
Hoopla — 8.65
Refreshments — 3.61
— 26.14

4 Luka works at a vet's surgery. Before a dog has an operation, Luka weighs it and records the mass to the nearest kilogram. Copy and complete the table, showing the upper and lower bound for the mass of each dog.

Dog	Recorded mass, kg	Lower bound, kg	Upper bound, kg
Murphy	32		
Kim	15		
Toby	12		
Dinky	26		

5 Demi is painting a wall. She measures the dimensions as 2.5 m by 6.3 m, to the nearest tenth of a metre. What is the upper bound for the area of the wall?

Hint

Use the upper bound of each measurement to calculate the area.

4.2 Estimating

One method for estimating the answer to a calculation is to round each number to one significant figure.

$$\frac{273 \times 49}{28} \approx \frac{300 \times 50}{30} = 500$$

$\approx$ means approximately equal to

Example 6

Q Niamh needs four tins of paint to decorate her flat. Each tin costs £38.99. Niamh has £150.

Use estimation to decide whether Niamh has enough money to buy the paint that she needs. Show your working.

A Paint costs approximately £40 per tin.

$4 \times £40 = £160$

Niamh does not have enough money to buy all the tins of paint.

Example 7

Q Estimate the answer to each calculation.

a $\dfrac{9.7 + 48}{0.32}$

b $\dfrac{306 \times 2.93}{0.051}$

A a $\dfrac{9.7 + 48}{0.32} \approx \dfrac{10 + 50}{0.3}$

$= \dfrac{60}{0.3}$

$= \dfrac{600}{3}$

$= 200$

b $\dfrac{306 \times 2.93}{0.051} \approx \dfrac{300 \times 3}{0.05}$

$= \dfrac{900}{0.05}$

$= \dfrac{90\,000}{5}$

$= 18\,000$

Exam tips Remember you can use estimation to check any of your calculations.

Practice questions

1 Round each number to one significant figure.

462   32   103   97   35.6

2 Rich rounded a number to one significant figure. The result was 60. Which of these could have been his original number?

| 57.6 | 64.9 | 53 | 60.4 | 54.9̇ |

3 Estimate by rounding: $\dfrac{12.8 + 66.2}{8.4 + 11.06}$

4 Estimate the value of $\dfrac{\sqrt{120.44}}{9.406}$.

12 Estimate the value of $\dfrac{4.3^3 - 22.4}{\sqrt{16.08}}$.

13 Estimate the value of $\dfrac{9.04^2 + 5.03}{0.53}$.

14 The answer to a question is given as 13 kg, correct to the nearest kilogram. Write down the upper and lower bounds of the answer.

15 A bag of toy stuffing contains 2 kg, correct to the nearest tenth of a kilogram. What is the upper bound for the mass of stuffing in the bag?

16 Johanna is knitting Poppy badges to sell for Remembrance Day, giving all the money to the Royal British Legion. The length of wool is 265 m, correct to the nearest 5 m. For each badge, she uses 35 m, correct to the nearest metre. She charges £4 per badge. What is the minimum amount of money she can expect to make from using this ball of wool?

17 Cara weighs herself on two different sets of scales. One set gives her mass as 58 kg, correct to the nearest kilogram. The other set gives her mass as 58 kg, correct to the nearest tenth of a kilogram. What are the upper and lower bounds of her mass?

18 A rectangle has length 6 cm and width 3 cm, each correct to the nearest centimetre. Work out the upper bound for the area of the rectangle.

19 Jack and Zeke work in a pub. One night, they are comparing the total amount of tips they each have. Jack says he has £12, correct to the nearest pound. Zeke says he has £12.40, correct to the nearest 10p. Which of them could have more money? Give a reason for your answer.

20 Dan needs to lay some turf. He measures the area to be covered as 7.8 m by 9.6 m. Both measurements are correct to the nearest 0.1 m.

Turf costs £3.55 per square metre. Work out the total cost of the turf Dan must buy to ensure he can cover the whole garden.

21 The speed, V, of a train is 72km/h, correct to the nearest km/h. Write down an inequality to show the error interval of V.

22 A distance, d, is given as 7.4m truncated to 1 decimal place. Complete the error interval for the distance, d.

Now go back to the list of objectives at the start of this chapter.
How confident do you now feel about each of them?

5 Formulae and expressions

Objectives

Before you start this chapter, mark how confident you feel about each of the statements below:

	▶	▶▶	▶▶▶
I can select an expression, equation, formula or identity from a list.			
I can simplify expressions by collecting like terms.			
I can multiply together two simple algebraic expressions.			
I can use the laws of indices when multiplying and dividing algebraic terms.			
I can substitute numbers into a word formula.			
I can substitute numbers into algebraic expressions.			
I can write expressions and formulae to solve problems, representing a situation.			

Check-in questions

- Complete these questions to assess how much you remember about each topic. Then mark your work using the answers at the back of the book.
- If you score well on all sections, you can go straight to the Revision checklist and Exam-style questions at the end of the chapter. If you don't score well, go to the chapter section indicated and work through the examples and practice questions there.

1 Say whether each statement is true or false.

    a $a \times b$ is the same as ab **b** $a + b$ is the same as ab

    c $a \times b$ is the same as ba **d** $a \div (b \times b)$ is the same as ab^2 Go to 5.1 ▶

2 Simplify these expressions.   Go to 5.2 ▶

    a $a + a + a + a$ **b** $5a + 2b + 3a - b$

    c $6a - 3b + 2a - 4b$ **d** $12xy + 4xy - xy$

    e $3a^2 - 6b^2 - 2b^2 + a^2$ **f** $5xy - 3yx + 2xy^2$

3 Simplify these expressions.   Go to 5.2 ▶

    a $5ab - 2bc + 6bc - 7ab$ **b** $d \times d \times d \times d$ **c** $5m \times 3n$

4 Simplify.   Go to 5.3 ▶

    a $2b^4 \times 3b^6$ **b** $8b^{-12} \div 4b^4$ **c** $(3b^4)^2$ **d** $(5x^2y^3)^{-2}$

5 If $a = \frac{3}{5}$ and $b = -2$, work out the value of each expression. Give your answers to three significant figures where appropriate.

Go to 5.4

a $ab - 5$ b $a^2 + b^2$ c $3a - 6ab$

6 Casey is C years old. Her daughter is 31 years younger than she is. The sum of their ages is 53. How old is Casey? How old is her daughter?

Go to 5.5

5.1 Algebraic terms and conventions

The table shows some words commonly used in algebra.

Word	Meaning	Example(s)
Term	A single number or letter or a collection of numbers and/or letters multiplied together.	$6a$, $2ab$, $3(x-1)$
Expression	An arrangement of terms joined by an operator, usually $+$ or $-$.	$2a + 3b - 4$
Equation	Two terms or expressions connected by an equals sign.	$x + 2 = 5$
Formula	A special type of equation that connects different variables; the value of one variable depends on the value(s) of the others.	$v = u + at$ (The value of v can be found when the values of u, a and t are known.)
Identity	An equation that is true for all numerical values of the letter variables. An $\equiv$ sign is used in place of $=$.	$2(x + 2) \equiv 2x + 4$ (This will always be true for any value of x.)
Function	A relationship between two sets of values, such that a value from the first set maps onto a value in the second set.	$y = 4x + 2$ The value of y can be calculated for any value of x.

Algebraic terms are usually written in a certain way, shown on the right.

Note that the number is usually written first and then the letters in alphabetical order, so $c \times a \times 5$ is written as $5ac$.

Term	Equivalent expression
$3c$	$3 \times c$ **or** $c \times 3$ **or** $c + c + c$
ab	$a \times b$ **or** $b \times a$
b^2	$b \times b$
$3b^2$	$3 \times b \times b$
$\dfrac{a}{2}$	$a \div 2$

Exam tips $3a^2$ is not the same as $(3a)^2$. $3a^2$ is 3 lots of a^2. $(3a)^2$ is 3 multiplied by a, then all of it squared.

Practice questions

1 Which expression below is equivalent to $(3a)^2$?

$3a$ $9a^2$ $3a^2$ $9a$ $6a^2$ $6a$

2 Match each expression on the left with the equivalent expression on the right.

$a \times b$ $\dfrac{ab}{2}$

$a \div b$ $2ab$

$a \times a$ ab

$a \times b \div 2$ a^2

$2 \times a \times b$ $\dfrac{a}{b}$

3 Which of these are equations?

| $y = 3x - 6$ $5x - 3y + 6$ $y \equiv 4x - 3$ $y = 3$ $y + 3x = -6$ |

4 Simplify each expression as far as possible.

  a $3 \times a$ **b** $a + b$ **c** $3 \times a + 6 \times b$ **d** $3 \times a \times b$

5 Match each expression on the left with the equivalent simplified expression on the right.

$2 \times 7 \times m$ $\dfrac{14}{m}$

$14 \div m$ $14 + m$

$6 + 8 + m$ $14m$

$2 + 7 \times m$ $2 + 7m$

5.2 Simplifying expressions

Collecting like terms

Expressions can be simplified by collecting **like terms**.

Like terms include the same letter combinations.

$$5ab - 3c - 6b^2 + 7$$

ab term c term b^2 term   number or constant term

Some terms can be simplified by **cancelling**. For example, $\dfrac{15x}{3} = 5x$ because $\dfrac{15}{3} = 5$.

Example 1	**Q** Simplify.
	a $5a + 3a$ **b** $3a - 4b + 2a + 3b$ **c** $5a^2 + 3a^2 - 2a^2$ **d** $2ab + 3ba$
	A **a** $8a$ **b** $5a - b$ **c** $6a^2$ **d** $5ab$

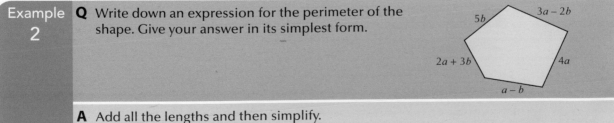

Example 2	**Q** Write down an expression for the perimeter of the shape. Give your answer in its simplest form.
	A Add all the lengths and then simplify.
	$5b + 3a - 2b + 4a + a - b + 2a + 3b = 10a + 5b$

Multiplying letters and numbers

Algebraic expressions that are multiplied together can often be simplified, e.g. $5a \times 2b = 10ab$.

<table>
<tr><td>Example
3</td><td>Q Simplify these expressions.

    a $3a \times 4b$ b $5a \times 3b \times 2c$ c $2a \times 3a$</td></tr>
<tr><td></td><td>A a $3a \times 4b = 3 \times 4 \times a \times b$ b $5a \times 3b \times 2c = 5 \times 3 \times 2 \times a \times b \times c$
           $= 12ab$ $= 30abc$

    c $2a \times 3a = 2 \times 3 \times a \times a$
          $= 6a^2$ $\boxed{\text{Remember } a \times a = a^2}$</td></tr>
</table>

Exam tips   Remember that x means $1x$. Remember also to use the order of operations.

Practice questions

1   Simplify each expression as far as possible.

    a $3x + 5x + 10x - 8x$ **b** $7y - 8y + 4y - 2$ **c** $a + a^2 + 5a - 6$

2   Simplify these expressions.

    a $5a \times 6a$ **b** $21x \times 3y$ **c** $5b^2 \times 16b$

3   Match each expression with the correct simplified version.

    $5a + 3b - 2a - 6b$ $6ab^2$

    $3a + 13b^2 - 16b^2$ $3a - 3b$

    $3b^2 \times 5a - 3a$ $3a - 3b^2$

    $3b^2 \times 2a$ $15ab^2 - 3a$

4   Write down an expression for the perimeter of the shape. Give your answer in its simplest form.

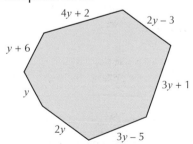

5   Write down an expression for the perimeter of the shape. Give your answer in its simplest form.

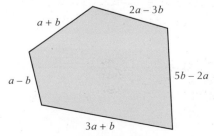

5.3 Laws of indices

The laws of indices (Chapter 1) are also used in algebra. The base must be the same for the laws of indices to apply.

$$a^n \times a^m = a^{n+m} \qquad a^n \div a^m = a^{n-m}$$
$$(a^n)^m = a^{n \times m} \qquad a^0 = 1$$
$$a^1 = a \qquad a^{-n} = \frac{1}{a^n}$$

Example 4

Q Simplify the following.

    a $5y^2 \times 3y^6$ **b** $a^4 \times a^{-6}$

A **a** $5y^2 \times 3y^6 = 5 \times 3 \times y^2 \times y^6$ •————— | Multiply the numbers and add the indices. |

       $= 15y^8$

  b $a^4 \times a^{-6} = a^{4-6}$

       $= a^{-2}$

       $= \dfrac{1}{a^2}$

Example 5

Q Simplify the following.

    a $(4x^3)^2$ **b** $(3x^4y^2)^3$ **c** $(2x)^{-3}$

A **a** $(4x^3)^2 = 16x^6$ •——— | Remember to square the 4 and multiply the indices. |

    If in doubt, write it out: $4x^3 \times 4x^3 = 16x^6$

  b $(3x^4y^2)^3 = 27x^{12}y^6$ •——— | $3x^4y^2 \times 3x^4y^2 \times 3x^4y^2 = 27x^{12}y^6$ |

  c $(2x)^{-3} = \dfrac{1}{(2x)^3}$

       $= \dfrac{1}{8x^3}$

Exam tips   Remember to use $a^{-n} = \dfrac{1}{a^n}$ when a question involves negative powers.

Example 6

Q Simplify these.

  a $\dfrac{15b^4 \times 3b^7}{5b^2}$ **b** $\dfrac{16a^2b^4}{4ab^3}$ **c** $\dfrac{9x^2y \times 2xy^3}{6xy}$

A **a** $\dfrac{15b^4 \times 3b^7}{5b^2} = \dfrac{45b^{11}}{5b^2}$ **b** $\dfrac{16a^2b^4}{4ab^3} = 4ab$ **c** $\dfrac{9x^2y \times 2xy^3}{6xy} = \dfrac{18x^3y^4}{6xy}$

         $= 9b^9$ $= 3x^2y^3$

Practice questions

1 Simplify each expression.

    a $a^7 \times a^3$ **b** $b^5 \times b^6$ **c** $c^3 \times c^4$ **d** $8d^3 \times 4d^2$

2 Simplify each expression.

    a $a^7 \div a^3$ **b** $b^{15} \div b^6$ **c** $c^3 \div c^4$ **d** $8d^5 \div 4d^2$

3 Say whether each statement is true or false.

    a $a^7 \times a^4$ is the same as a^{28}

    b $b^{20} \div b^2$ is the same as b^{10}

    c $\dfrac{5y^3 \times 6y^7}{10y^4}$ is the same as $3y^{17}$

    d $7p^{-2}$ is the same as $\dfrac{7}{p^2}$

4 Match each expression on the left with the correct simplified version.

    $\dfrac{21a^3b^4}{7ab^2}$ $\dfrac{2b^3}{a^2}$

    $\dfrac{5a^2 \times b^2}{b}$ $\dfrac{4a^2}{b^2}$

    $\dfrac{16a^5 \times 2ab^3}{8a^4b^5}$ $5a^2b$

    $\dfrac{3ab^5 \times 4a}{6a^4b^2}$ $3a^2b^2$

5 Simplify $(4x^3y^5)^2$.

5.4 Substituting into formulae

Replacing a letter with a number is called **substitution**. Write out the expression first and then replace the letters with the values given.

Remember to use the order of operations (**BIDMAS**) when you work out the value (see Chapter 1).

Example 7	**Q** $a = 3b - 4c$
	Work out the value of a when $b = 4$ and $c = -2$.

A $a = (3 \times 4) - (4 \times -2)$

    $= 12 - (-8)$ ●——— Subtracting a negative number is the same as adding a positive number (see Chapter 1).

    $= 20$

Example 8

Q $v = u + at$

Work out the value of v when $u = 22$, $a = -2$ and $t = 6$.

A $v = 22 + (-2 \times 6)$

$ = 22 + (-12)$ ← Adding a negative number is the same as subtracting a positive number (see Chapter 1).

$ = 10$

Example 9

Q $S = kt^2$

Work out the value of S when $k = 4$ and $t = -3$.

A $S = 4 \times (-3)^2$

$ = 4 \times 9$ ← Remember that squaring a negative number gives a positive result (see Chapter 1).

$ = 36$

Example 10

Q The cost of hiring a car can be worked out using this rule: Cost = £85 + 48p per mile.

  a Josh hires a car and drives 130 miles. Work out the cost.

  b On a different day Josh pays £131.56 for hiring a car. How many miles did he drive?

A **a** $C = £85 + 48p$ per mile ← Change 48p into pounds (£0.48)

$ = 85 + 0.48 \times 130$

$ = 85 + 62.40$

$ = £147.40$

  b $131.56 = 85 + 0.48 \times$ number of miles

$131.56 - 85 = 0.48 \times$ number of miles

$46.56 = 0.48 \times$ number of miles

$46.56 \div 0.48 = 97$ miles ← Divide by the cost per mile to find the number of miles.

Example 11

Q A person's body mass index (BMI), b, is calculated using the formula $b = \dfrac{m}{h^2}$ where m is body mass in kilograms and h is height in metres.

A person is classed as overweight if their BMI is greater than 25.

Dan has height 179 cm and mass 84.5 kg. Is Dan classed as overweight? Show your working.

A $b = \dfrac{m}{h^2}$ $m = 84.5$ kg $h = 1.79$ m

$b = \dfrac{84.5}{1.79^2}$

$b = 26.37$

Yes, Dan is classed as overweight.

Exam tips Take extra care when you substitute negative into squared terms of a formula.

Practice questions

1 $y = 3x + 2$

Find the value of y for each of these values of x.

a 2   **b** 6   **c** 21   **d** –4

2 A hotel charges £2 to connect to the internet and then £1.50 for each hour. William uses the hotel internet for his meetings. How much is he charged for internet use for:

a 2 hours   **b** 4 hours   **c** 5 hours   **d** 8 hours?

3 Alicia sells cakes. She charges using this formula:

Cost in £ = $m \times 2 + 2.50$, where m is dependent on the type of cake.

Work out the cost of each of these cakes.

a Vanilla sponge, $m = 1$ **b** Chocolate sponge, $m = 1.5$

c Lemon drizzle, $m = 1.8$ **d** Carrot cake, $m = 2.1$

e Chocolate waterfall, $m = 2.75$

4 Shaun makes document deliveries by bike for law firms around a city centre. He uses this formula to calculate his charge:

Cost in £ = $1.2m + 0.5t + 3$, where m is the number of miles and t is the time in minutes.

Use the formula to calculate his charges for these deliveries.

	m	t
a	1.2	8
b	0.4	5
c	3.6	14
d	5.1	20

5 The cost of hiring a van is found by using the formula:

$C = 38d + 0.53m + 50$, where C is the cost in £, d is the number of days for which the van is hired and m is the number of miles driven.

Dylan hires a van for 3 days and drives 502 miles. Calculate the cost.

5.5 Writing formulae

$b = a + 6$ is a **formula**. The value of b is dependent on the value of a.

You need to be able to write a formula using information given.

Example 12

Q My brother is three years older than me. My mother is three times as old as me.

   a If I am n years old, write expressions for my brother's and mother's ages.

   b Write down a formula for the sum, s, of our ages.

A a If I am 12, my brother is $12 + 3 = 15$.   [Try using numbers first.]

     So if I am n years old, my brother is $n + 3$ years old.

     If I am 12, my mother is $3 \times 12 = 36$.

     So if I am n years old, my mother is $3 \times n$ or $3n$ years old.

   b $s = n + n + 3 + 3n$

     $s = 5n + 3$

Example 13

Q Lauren buys *x* books costing £7 each and *y* magazines costing 98p each. Write down a formula for the total cost, *T*, of the books and magazines Lauren buys.

A First, look at the units for the costs. They are different, so you need to write the cost of a magazine in pounds or the cost of a book in pence.

$T = 7x + 0.98y$ where *T* is cost in £   or $T = 700x + 98y$ where *T* is cost in pence

Exam tips   Always check that the units are the same for each term when you write a formula. The question will sometimes state the units required. Always make sure you write down the units with your answer if they are not given.

Example 14

Q Adult cinema tickets cost £*x* and child cinema tickets cost £*y*. Mr Khan buys 2 adult tickets and 4 child tickets. Write down a formula in terms of *x* and *y* for the total cost (£*C*) of the tickets.

A $C = 2x + 4y$

Practice questions

1 An isosceles triangle has angles of $2x$, $2x$ and $6x$.

   a Form an equation in terms of *x*.

   b Solve your equation to work out the value of *x*.

   c Calculate the sizes of the angles in the isosceles triangle.

2 Tom thought of a number, *n*. He divided it by 2, and then added 4. The result was one more than the number he first thought of.

   a Use the above information to set up an equation.

   b Solve your equation to work out the value of *n*.

3 Nigethan drew two rectangles. They both had the same area.

 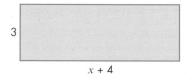

   a Set up an equation in terms of *x*.

   b Solve your equation to work out the value of *x*.

4 Kate is *K* years old. Kate's cousin is twice her age. The sum of their ages is 27. How old is Kate?

5 Paulo thought of a number. He multiplied his number by 4. He then added 3 to the result. He doubled that result and got 54.

   What number did he start with?

REVISION CHECKLIST

- Expressions are made up of a two or more terms joined by one or more operators (+, −).
- You can simplify an expression by collecting like terms. You can only collect together terms that include exactly the same letter combinations.
- Substitution is replacing letters in a formula with numbers.
- The laws of indices are:

$$a^n \times a^m = a^{n+m} \qquad a^n \div a^m = a^{n-m}$$

$$(a^n)^m = a^{n \times m} \qquad a^0 = 1$$

$$a^1 = a \qquad a^{-n} = \frac{1}{a^n}$$

- A negative power is the reciprocal of the positive power.

Exam-style questions

1   Write an expression for the perimeter of this rectangle.

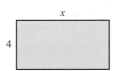

2   Write an expression for the perimeter of this triangle.

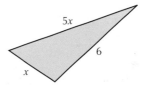

3   How many terms are there in this expression? $3x + 5b - 2c$

4   State whether each of these is an expression, an equation, a formula or an identity.

   a $5y - 7 = P$ **b** $2x + 6 - v$ **c** $y \equiv 4t + 6p - 7$ **d** $2(x + 3) = 15$

5   Simplify fully $5x \times 3y + 3 \times y^2$.

6   Write down an expression for the perimeter of this shape. Give your answer in its simplest form.

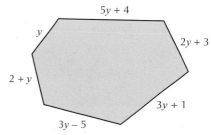

7   Bev and Mike are looking at the expression on the right: $5a - 3b$

   Bev says it cannot be simplified.

   Mike says it will simplify to $2ab$.

   Which of them is correct? Give a reason for your answer.

8 Write down an expression for the perimeter of the shape. Give your answer in its simplest form.

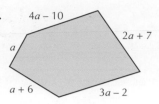

9 These two rectangles have the same perimeter. Calculate the value of x.

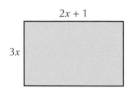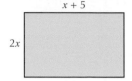

10 This triangle has an area of 40 cm². Work out the value of x.

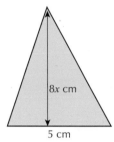

11 Marin is M years old. Her mum is 35 years older than her. The sum of their ages is 63 years. How old is Marin?

12 Arthur is A years old. His granddad is 68 years older than him. He calculates that half of the sum of their ages is 42 years. How old is Arthur?

13 Write as a single power. $a^9 \times a^6$

14 Write as a single power of b. $b^3 \div b^{10}$

15 Write as a single power of c. $c^3 \div c^8 \times c^6$

16 Work out the value of $P = mgh$ when $m = 20$, $g = 9.8$, $h = 11$.

17 Simplify fully: $\dfrac{20a^5b^4}{5ab^3}$

18 Hamish draws a regular hexagon and a regular pentagon with the same perimeter. The side length of the hexagon is $2x + 1$ and the side length of the pentagon is $3x - 2$.

  a Write down an equation in terms of x.

  b Use your equation to work out the side length of the pentagon.

19 a is an odd number and b is an even number. Say whether each statement is always true, sometimes true or never true.

  a ab = even   **b** $a + b$ = even   **c** b^2 = even   **d** ab^2 = even

20 A square and a rectangle have the same area. The side length of the square is $3x$ and the dimensions of the rectangle are 5 and $2x + 1$.

  Work out the area of the square.

Now go back to the list of objectives at the start of this chapter. How confident do you now feel about each of them?

6 Algebraic manipulation

Check-in questions

- Complete these questions to assess how much you remember about each topic. Then mark your work using the answers at the back of the book.

- If you score well on all sections, you can go straight to the Revision checklist and Exam-style questions at the end of the chapter. If you don't score well, go to the chapter section indicated and work through the examples and practice questions there.

1 Expand and simplify these expressions.

    a $t(3t - 4)$ **b** $4(2x - 1) - 2(x - 4)$ Go to 6.1

2 Expand and simplify these expressions.

    a $(x + 3)(x - 2)$ **b** $4x(x - 3)$ **c** $(x - 3)^2$ Go to 6.2

3 Factorise these expressions.

    a $12xy - 6x^2$ **b** $3a^2b + 6ab^2$ **c** $x^2 + 4x + 4$

    d $x^2 - 4x - 5$ **e** $x^2 - 100$ Go to 6.3

4 Factorise these expressions. Expand and simplify first if necessary.

    a $y^2 + y$ **b** $5p^2q - 10pq^2$

    c $(a + b)^2 + 4(a + b)$ **d** $x^2 - 5x + 6$ Go to 6.2, 6.3

5 Make u the subject of the formula $v^2 = u^2 + 2as$.   Go to 6.4

6.1 Expanding single brackets

Multiplying out brackets helps to simplify algebraic expressions.

The term outside the bracket is multiplied by each of the terms inside the bracket.

$$5(x + 6) = 5x + 30$$

Example 1	**Q** Expand and simplify these expressions. **a** $-2(2x + 4)$ **b** $5(2x - 3)$ **c** $8(x + 3) + 2(x - 1)$ **d** $3(2x - 5) - 2(x - 3)$

A **a** $-2(2x + 4) = -4x - 8$

   b $5(2x - 3) = 10x - 15$

   c $8(x + 3) + 2(x - 1) = 8x + 24 + 2x - 2$

                       $= 10x + 22$ **Collect like terms.**

   d $3(2x - 5) - 2(x - 3) = 6x - 15 - 2x + 6$

                        $= 4x - 9$ **Collect like terms.**

Exam tips   When you multiply out brackets, check that the signs are correct for each term. If the signs are the same, it is +. If the signs are different, it is −.

Practice questions

1   Expand and simplify.

   a $3(x + 5)$ **b** $7(y - 2)$ **c** $13(z + 5)$

2   Expand and simplify.

   a $-2(x + 4)$ **b** $2(2x + 7)$ **c** $-3(5x + 3y)$ **d** $5x(x + 4)$

3   Expand and simplify.

   a $2(x + 4) + 3(x + 1)$ **b** $3(y - 2) + 3(y + 3)$

   c $4(m + 5) - 2(m + 6)$ **d** $3(2m + 3) - 2(4m - 12)$

4   Deepak expands $3(4f - 2)$. His result is $7f - 2$.

   Identify two errors that he has made.

Hint

Expand both sides and show that they give the same result.

5   Show that $3(y + 2) - 2(y - 3) \equiv 2(y + 6) - y$.

This sign means 'is equivalent to'.

6.2 Expanding two pairs of brackets

Every term in the second bracket must be multiplied by every term in the first bracket.

Often, but not always, the two middle terms are like terms and can be collected together.

$$(x + 4)(x + 2) = x^2 + 2x + 4x + 8$$

$$= x^2 + 6x + 8$$

Example 2

Q Expand and simplify these expressions.

a $(x + 4)(2x - 5)$ **b** $(2x + 1)^2$ **c** $(3x - 1)(x - 2)$

d $(x - 4)(3x + 1)$ **e** $(2x + 3y)(x - 2y)$

A **a** $(x + 4)(2x - 5) = 2x^2 - 5x + 8x - 20$

$= 2x^2 + 3x - 20$

b $(2x + 1)^2 = (2x + 1)(2x + 1)$

$= 4x^2 + 2x + 2x + 1$

$= 4x^2 + 4x + 1$ •————— Remember that x^2 means x multiplied by itself.

c $(3x - 1)(x - 2) = 3x^2 - 6x - x + 2$

$= 3x^2 - 7x + 2$

d $(x - 4)(3x + 1) = 3x^2 + x - 12x - 4$

$= 3x^2 - 11x - 4$

e $(2x + 3y)(x - 2y) = 2x^2 - 4xy + 3xy - 6y^2$

$= 2x^2 - xy - 6y^2$

Exam tips   Remember to check that the signs are correct for each term.

Practice questions

1 Expand and simplify.

a $(y + 2)(y + 3)$ **b** $(x + 4)(x + 2)$ **c** $(m + 3)(m + 5)$

2 Expand and simplify.

a $(x + 4)(x - 1)$ **b** $(y + 6)(y - 2)$ **c** $(m - 2)(m + 8)$ **d** $(x + 5)(x - 5)$

3 Expand and simplify.

a $(3x + 2)(x + 4)$ **b** $(5y - 2)(5y + 4)$ **c** $(7 - 2y)(3 - 8y)$

4 Match the expression on the left with the correct expansion on the right.

$(x - 2)(x + 2)$	$2x^2 - x - 1$
$(x + 3)(x - 1)$	$x^2 - 4$
$(x + 4)(x - 2)$	$x^2 + 2x - 3$
$(2x + 1)(x - 1)$	$x^2 + 2x - 8$

5 Tash expands and simplifies $(x + 2)(x - 3)$. Her result is $x^2 - 1$. Identify any errors she has made.

6.3 Factorisation

Factorisation is the opposite of expansion. It involves looking for common factors and putting them outside brackets.

One pair of brackets

To factorise $4x + 6$:

- Identify the HCF of the terms. 2
- Divide the expression by the HCF. $2x + 3$
- Write the HCF outside the brackets and the rest of the expression inside the brackets. $2(2x + 3)$

When it is expanded or multiplied, it will equal the original expression.

Example 3	**Q** Factorise fully.
	a $8x - 16$ **b** $3x + 18$ **c** $5x^2 + x$ **d** $4x^2 + 8x$
	A a $8(x - 2)$ **b** $3(x + 6)$ **c** $x(5x + 1)$ **d** $4x(x + 2)$

Exam tips Remember to check that you have put all the common factors outside the bracket, otherwise the expression is not fully factorised.
Check there is no common factor inside the bracket for your final answer.

Two pairs of brackets

Two pairs of brackets are obtained when a quadratic expression of the type $ax^2 + bx + c$ is factorised.

Example 4	**Q** Factorise these expressions.
	a $x^2 + 4x + 3$ **b** $x^2 - 7x + 12$ **c** $x^2 + 3x - 10$
	A a Find two numbers with a product of 3 (the constant) and a sum of 4 (the coefficient of x): 3 and 1 ($3 \times 1 = 3$ and $3 + 1 = 4$)
	$(x + 1)(x + 3)$
	b Find two numbers with a product of 12 (the constant) and a sum of -7 (the coefficient of x): -4 and -3 ($-4 \times -3 = 12$ and $-4 + -3 = -7$)
	$(x - 3)(x - 4)$
	c Find two numbers with a product of -10 (the constant) and a sum of 3 (the coefficient of x): -2 and 5 ($-2 \times 5 = -10$ and $-2 + 5 = 3$)
	$(x + 5)(x - 2)$ —————— Note that although $-5 \times 2 = -10$, $-5 + 2 = -3$

Example
5

Q Factorise these expressions.

   a $x^2 - 64$ **b** $81x^2 - 25y^2$

A **a** $x^2 - 64 = (x + 8)(x - 8)$ •

    b $81x^2 - 25y^2 = (9x + 5y)(9x - 5y)$

> This is known as the 'difference of two squares'. In general, $x^2 - a^2 = (x + a)(x - a)$.

Practice questions

1 Factorise these expressions.

   a $4x + 10$ **b** $15x + 35$ **c** $14x - 120$ **d** $6x - 30$

2 Factorise fully these expressions.

   a $5y + 10x + 25z$ **b** $16x - 48x^2 + 56xz$ **c** $14x - 42y + 63z$ **d** $-9m + 12n - 21p$

3 Factorise these quadratic expressions.

   a $x^2 + 5x + 6$ **b** $x^2 + 10x + 16$ **c** $x^2 - 10x + 21$ **d** $x^2 + 6x - 16$

4 Factorise these expressions.

   a $x^2 - 9$ **b** $y^2 - 49$ **c** $m^2 - 1$ **d** $9t^2 - 121y^2$

5 In an exam, Ravjeet is asked to fully factorise $25x^2 - 40x + 15xy$.

   As her answer, she writes $5(5x^2 - 8x + 3xy)$.

   She does not score full marks. Explain why.

6.4 Rearranging formulae

The subject of a formula is the letter that appears on its own on one side of the formula.

Any letter in a formula can become the subject if you rearrange the formula. It is important to do the same things to both sides of the formula as you rearrange it.

Example
6

Q Make c the subject of the formula: $b = c - a$

A $b = c - a$ •

   $b + a = c$

   So $c = b + a$

> Add a to both sides.

Example
7

Q Make a the subject of the formula: $b = (a - 3)^2$

A $b = (a - 3)^2$

   $\pm\sqrt{b} = a - 3$ •

   $\pm\sqrt{b} + 3 = a$ •

   $a = \pm\sqrt{b} + 3$

> Deal with the power first. Square root both sides of the formula. There will be a positive and negative solution, shown as ± (see Chapter 1).

> Add 3 to both sides of the formula.

When the subject occurs on both sides of the equals sign, they need to be collected on one side. It is easier to collect the terms involving the new subject on the side where there are more of them.

Example 8

Q Make x the subject of the formula: $5(y + x) = 8x + 3$

A $5(y + x) = 8x + 3$

$5y + 5x = 8x + 3$

$5y - 3 = 8x - 5x$ ● — Collect the x terms on one side.

$5y - 3 = 3x$

$x = \dfrac{5y - 3}{3}$ ● — Divide both sides by 3 to make x the subject.

Exam tips Show all steps in your working. This will help you gain marks for your method even if your final answer is incorrect.

Practice questions

1 Rearrange each formula to make a the subject.

a $x = a + 6$ **b** $y = a - 6$ **c** $z = 6 - a$

2 Rearrange each formula to make b the subject.

a $x = 4b + 6$ **b** $y = 5b - 3$ **c** $z = 6 - 2b$

3 Rearrange each formula. The subject is given in brackets.

a $V = IR$ (R) **b** $A = \dfrac{bh}{2}$ (b) **c** $P = mgh$ (g)

4 Make y the subject of the formula. $5y + m = 4(y - 3)$

REVISION CHECKLIST

● To expand single brackets, multiply the term outside the brackets by everything inside the brackets.

● To expand two pairs of brackets, multiply each term in the second bracket by each term in the first bracket.

● Factorising is the opposite of expanding brackets. Look for the highest common factor (HCF).

● You can rearrange a formula to make any letter the subject. When rearranging a formula, you must do the same thing to both sides.

Exam-style questions

1 Which of these are true?

a $5(a + 3) = 5a + 3$ **b** $3(b + 2) = 3b + 6$

c $5c(c - 2) = 5c^2 - 2c$ **d** $2d(d + 4) = 2d^2 + 8d$

2 Factorise this expression. $3pr + 12qr - 21r^2$

3 Ted factorises $5x - 40$. His answer is $5(x - 40)$.

Show, by expanding his answer, that he is incorrect.

4 Expand and simplify this expression. $\quad 4(x + 3) + 2(x + 1)$

5 Expand and simplify. $\quad (x + 8)(2x - 3)$

6 Expand and simplify. $\quad (2y + 6)(3y - 2)$

7 Write an expression for the perimeter of the rectangle. Fully expand and simplify your answer.

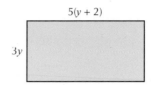

$5(y + 2)$

$3y$

8 Write an expression for the area of this square. Fully expand and simplify your answer.

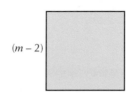

$(m - 2)$

9 The area of this rectangle is $8m + 16$. Write an expression for the length of the rectangle if the width is 8.

8 | $8m + 16$

10 Rearrange this formula to make m the subject. $\quad y = mx + c$

11 Rearrange this formula to make p the subject. $\quad W = \dfrac{pr}{q}$

12 Expand and simplify this quadratic expression. $\quad (2x + 3)(5x - 4)$

13 Show that $3(x + 4) + 2(x - 3) \equiv 5(x + 1) + 1$.

14 Factorise this expression. $\quad x^2 - 144$

15 Factorise. $\quad 9p^2 - 25q^2$

16 Rearrange this formula to make F the subject. $\quad \dfrac{5(F - 32)}{9} = C$

17 Show that $(2x - 4y)(2x + 4y) = 4x^2 - 16y^2$.

18 Rearrange this formula to make r the subject. $\quad S = 4\pi r^2$

19 Rearrange this formula to make r the subject. $\quad V = \dfrac{4}{3}\pi r^2$

20 Factorise this expression. $\quad x^2 - x - 20$

**Now go back to the list of objectives at the start of this chapter.
How confident do you now feel about each of them?**

7 Equations and inequalities

Objectives

Before you start this chapter, mark how confident you feel about each of the statements below:	▶	▶▶	▶▶▶
I can solve simple linear equations.			
I can solve linear equations with the unknown appearing on both sides of the equation.			
I can solve linear equations that contain brackets.			
I can solve equation problems.			
I can write whole number values that satisfy an inequality.			
I can solve linear inequalities and show the results on a number line.			
Stretch objectives Go to www.collins.co.uk/edexcelpost16			
I can solve linear equations involving fractions.			
I can solve quadratic equations by factorising.			
I can solve two inequalities and compare them to find values that satisfy both inequalities.			

Check-in questions

- Complete these questions to assess how much you remember about each topic. Then mark your work using the answers at the back of the book.

- If you score well on all sections, you can go straight to the Revision checklist and Exam-style questions at the end of the chapter. If you don't score well, go to the chapter section indicated and work through the examples and practice questions there.

1. Solve these equations.   Go to 7.1

   a $2x - 6 = 10$ **b** $5 - 3x = 20$

2. Solve these equations.   Go to 7.2

   a $6x + 3 = 2x - 10$ **b** $7x - 4 = 3x - 6$

3. Solve these equations.   Go to 7.3

   a $4(2 - 2x) = 12$ **b** $5(x + 1) = 3(2x - 4)$ **c** $5 - 2x = 3(x + 2)$

④ The perimeter of this triangle is 40 cm. Work out the value of x and the length of the shortest side.

Go to 7.4

$2x - 5$ cm

$2x + 6$ cm

$4x - 5$ cm

⑤ n is an integer such that $-6 < 2n \leq 8$

List all the possible values of n.

Go to 7.5

7.1 Linear equations of the form $ax + b = c$

An **equation** usually involves an unknown value that can be calculated. When you solve an equation, you find the unknown quantity. The solution to $5x + 2 = 12$ is $x = 2$.

The left and right sides of an equation are the same – they balance. To keep them in balance, whatever you do to one side of the equation (for example, adding 3), you also need to do to the other side.

Example 1	**Q** Solve these equations.
	a $3x = 15$
	b $\frac{x}{3} = 6$

A **a** $3x = 15$

$x = \dfrac{15}{3}$ ● ─ Divide both sides by 3.

$x = 5$

b $\dfrac{x}{3} = 6$

$x = 6 \times 3$ ● ─ Multiply both sides by 3.

$x = 18$

Sometimes, you will need more than one step to solve an equation.

Example
2

Q Solve these equations.

   a $5x - 2 = 13$ **b** $3x + 1 = 13$ **c** $\frac{x}{6} - 1 = 3$

A **a** $5x - 2 = 13$

      $5x = 13 + 2$ •————⟨ Add 2 to both sides. ⟩

      $5x = 15$

      $x = \dfrac{15}{5}$ •————⟨ Divide both sides by 5. ⟩

      $x = 3$

  b $3x + 1 = 13$

      $3x = 13 - 1$ •————⟨ Subtract 1 from both sides. ⟩

      $3x = 12$

      $x = \dfrac{12}{3}$ •————⟨ Divide both sides by 3. ⟩

      $x = 4$

  c $\frac{x}{6} - 1 = 3$

      $\dfrac{x}{6} = 3 + 1$ •————⟨ Add 1 to both sides. ⟩

      $\dfrac{x}{6} = 4$

      $x = 4 \times 6$ •————⟨ Multiply both sides by 6. ⟩

      $x = 24$

Exam tips   Always check your answer by substituting it back into the original equation.

Practice questions

1   Solve these equations.

   a $5x = 75$ **b** $6x = 42$ **c** $\frac{x}{7} = 7$ **d** $\frac{x}{3} = 18$

2   Solve these equations.

   a $2x + 5 = 21$ **b** $3x - 2 = 13$ **c** $6x - 4 = 32$ **d** $5x + 7 = 47$

3   Solve these equations.

   a $\frac{x}{7} + 3 = 10$ **b** $\frac{x}{4} - 6 = 0$ **c** $\frac{3x}{5} = 6$ **d** $\frac{y - 6}{4} = 1$

4   Match the equation to the correct solution.

   $3x = 24$ $x = -2$

   $5x + 2 = 22$ $x = 8$

   $\frac{x}{2} + 18 = 38$ $x = 4$

   $6x - 9 = -21$ $x = 40$

5   One of these equations has a different solution to the others. Which one?

   a $3x + 6 = 15$ **b** $6x - 7 = 8$ **c** $\frac{10}{x} + 4 = 8$ **d** $\frac{25 - 2x}{5} = 4$

7.2 Linear equations of the form $ax + b = cx + d$

Q Solve: $7x - 4 = 3x + 8$

A $7x = 3x + 12$ — Add 4 to both sides.

$4x = 12$ — Subtract $3x$ from both sides.

$x = \dfrac{12}{4}$ — Divide both sides by 4.

$x = 3$

Check by substituting $x = 3$ into both sides of the equation:

$7 \times 3 - 4 = 17$

$3 \times 3 + 8 = 17$

Since both the left-hand side of the equation and the right-hand side of the equation give the same answer, $x = 3$ is correct.

Q Solve: $5x + 3 = 2x - 5$

A $5x = 2x - 5 - 3$ — Subtract 3 from both sides.

$5x = 2x - 8$

$5x - 2x = -8$ — Subtract $2x$ from both sides.

$3x = -8$

$x = -\dfrac{8}{3}$ — Divide both sides by 3.

$x = -2\dfrac{2}{3}$

Exam tips Take the unknown (x) to the side where there are more of them. Always check your answer by substituting it back into the original equation.

Practice questions

1 Solve these equations.

  a $6x + 2 = 8x$ **b** $3x - 3 = 2x$ **c** $5x + 7 = 10x - 8$

2 Solve these equations.

  a $2x + 3 = 6x - 13$ **b** $9x - 7 = 4x - 12$ **c** $8x - 6 = 2x + 3$

3 Cliona and Siobhan are playing 'think of a number'. Cliona says she multiplies by 5 then adds 4. Siobhan multiplies by 3 and then adds 10. They both think of the same starting number and get to the same result. What was the starting number?

4 Solve these equations.

  a $5x + 13 = 3x + 15$ **b** $5 - 4x = 15 + x$ **c** $7 - 4x = 16 + 2x$

7.3 Linear equations with brackets

Example 5

Q Solve: $5(x-1) = 3(x+2)$

A $5x - 5 = 3x + 6$ — Multiply out the brackets first.

$5x = 3x + 11$

$2x = 11$

$x = \dfrac{11}{2}$

$x = 5.5$

Example 6

Q Solve: $5(2x+3) = 2(x-6)$

A $10x + 15 = 2x - 12$

$10x = 2x - 12 - 15$

$10x = 2x - 27$

$10x - 2x = -27$

$8x = -27$

$x = -\dfrac{27}{8}$

$x = -3\frac{3}{8}$

Example 7

Q Joe is asked to solve the equation $3(x-6) = 42$.

Here is his working:

$3(x - 6) = 42$
$3x - 6 = 42$
$3x = 42 + 6$
$3x = 48$
$x = 16$

What mistake did he make?

A Joe forgot to multiply the 3 and the –6.

When he multiplied out the bracket, the result should have been $3x - 18 = 42$.

Exam tips The solution to an equation can be a positive or negative integer, fraction or decimal. Remember to write the exact value of your answer unless you are asked to write it to a given degree of accuracy.

Practice questions

1 Solve these equations.

a $3(x + 2) = 21$ **b** $4(x - 2) = 4$ **c** $7(x + 7) = 70$

2 Solve these equations.

a $4(2x + 3) = 100$ **b** $6(3x - 2) = 150$ **c** $2(5x + 3) = 15x + 6$ **d** $3(x - 4) = -3x$

3 Solve these equations.

a $5x + 13 = 3(x + 5)$ **b** $2(1 - 5x) = 3(5x - 1)$

c $5(2x - 3) = 2(x + 2) + 2(2x - 1) - 7$ **d** $3(2x - 1) + 4(x + 3) = 5(2x - 1) + 4(3x - 1)$

4 A square has side length $(2x + 3)$. The perimeter is 60 cm. Find the length of one side in centimetres.

5 The solution for $a(bx + 7) = c$ is $x = 10$. Give possible values for a, b and c.

7.4 Equation problems

When solving equation problems, the first step is to write down the information that you know.

Example 8	**Q** The perimeter (distance around the outside edge) of this rectangle is 30 cm. Work out the value of y and the length of the longest side of the rectangle.	$3y + 4$ cm $2y$ cm

A $3y + 4 + 2y + 3y + 4 + 2y = 30$ ———— Write down what you know.

$10y + 8 = 30$ ———— Simplify the expression and then solve.

$10y = 30 - 8$

$10y = 22$

$y = 2.2$

Length of longest side: $3 \times 2.2 + 4 = 10.6$ cm

Example 9

Q The diagram shows a quadrilateral with the sizes of its angles marked.

Work out the size of the smallest angle.

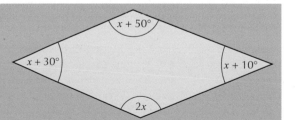

A $x + 30 + x + 50 + x + 10 + 2x = 360$ ———— The angles in a quadrilateral add up to 360°.

$5x + 90 = 360$

$5x = 360 - 90$

$5x = 270$

$x = 54$

So, smallest angle: $x + 10° = 64°$ ———— Check that you have worked out what was asked for in the question.

Practice questions

1 The perimeter of this triangle is 49 cm. Find the value of x.

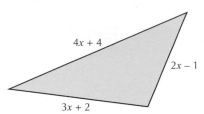

2 The angles of a triangle are $4y$, $7y + 3°$ and $2y + 8°$. Calculate the size of the largest angle.

3 The square and the rectangle have the same perimeter. Calculate the value of m.

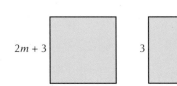

4 The perimeter of the square and triangle are equal. Calculate the value of x.

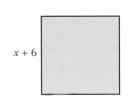
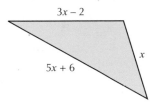

5 Look at this quadrilateral. Work out the size of the largest angle.

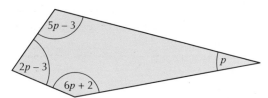

6 The mean of $3x$, $4x$, $2x + 1$ and $x - 3$ is 8. Find the value of x.

7.5 Inequalities

An **inequality** is a statement comparing two quantities that are not equal.

The inequality symbols

> means **greater than**

< means **less than**

⩾ means **greater than or equal to**

⩽ means **less than or equal to**

Solving inequalities

Inequalities can be solved in a similar way to equations. The only difference is that when you multiply or divide by a negative number, you must reverse the inequality sign.

Example 10

Q Solve: $4x - 2 \geqslant 6$

A $4x \geqslant 6 + 2$ ⟶ Add 2 to both sides.

$4x \geqslant 8$

$x \geqslant \dfrac{8}{4}$ ⟶ Divide both sides by 4.

$x \geqslant 2$

Example 11

Q Solve: $2x - 2 < 10$

A $2x < 10 + 2$

$2x < 12$

$x < 6$

Example 12

Q Solve: $3 - 2x \geqslant 9$

A $-2x \geqslant 9 - 3$

$-2x \geqslant 6$

$x \leqslant \dfrac{6}{-2}$ ⟶ Divide by −2 and reverse the inequality sign.

$x \leqslant -3$

Exam tips Check that your answer contains the correct inequality sign and you have not accidentally replaced it with an '=' sign.

Number lines

Inequalities can be shown on a number line.

$x < 6$

An open circle means that the number is not included.

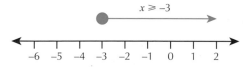

$x \geqslant -3$

A solid circle means that the number is included.

$-4 \leqslant x < 2$

The integer values that satisfy this inequality are −4, −3, −2, −1, 0, 1 and 2.

Practice questions

1 Represent these inequalities on a number line.

   a $x < 7$ **b** $x > 5$ **c** $x \leqslant 4$ **d** $5 \leqslant x < 9$

2 Write the inequality represented by each number line.

a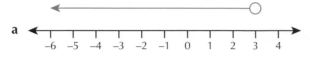

b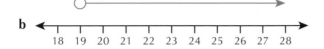

c

d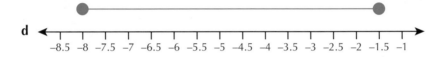

3 Solve these inequalities.

   a $2x + 3 > 15$ **b** $3x - 6 < 54$ **c** $7x + 1 \leqslant 57$ **d** $5x - 6 \geqslant 19$

REVISION CHECKLIST

- When solving an equation:
  - Do the same things (operations) to both sides.
  - Write out all the steps in your solution.
  - Check by substituting in your answer.

- To solve an equation problem, write down what you know in equation form, simplify and then solve.

- The inequality symbols are:
  - $>$ greater than
  - $<$ less than
  - $\geqslant$ greater than or equal to
  - $\leqslant$ less than or equal to

- When you represent an inequality on a number line, an open circle means the value is not included in the inequality and a solid circle means the value is included in the inequality.

Exam-style questions

1　Solve: $3x - 4 = -19$

2　A triangle has angles $3x$, $2x + 2°$ and $4x - 2°$.
Find the size of the smallest angle.

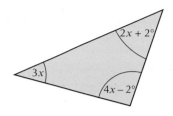

3　Choose the correct solution for $6x - 6 = 18$.

| $x = 12$ | $x = 24$ | $x = 2$ | $x = 4$ | $x = 6$ | $x = 3$ |

4　Dennis writes out his solution to $5x - 4 = 24$ as:

When he checks his answer, he finds that
$5 \times 4 - 4 = 16$. Explain where he has gone
wrong and find the correct value for x.

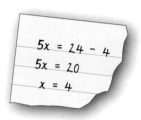

$5x = 24 - 4$
$5x = 20$
$x = 4$

5　Write the inequality shown by this number line.

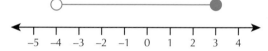

6　A square has sides of length $2x - 2$ and $x + 4$.
　　a Explain why $2x - 2 = x + 4$.
　　b Calculate the perimeter of the square.

$2x - 2$

$x + 4$

7　Solve: $3x + 7 = 5x + 2$

8　Solve: $4(x - 7) = x - 1$

9　Which equation does not have the same solution as the others?
　　$x + 1 = 6$　　　$6x = 36$　　　$6 - x = 1$　　　$5x = 25$

10　Match each equation with the correct solution.

　　$3x = 21$　　　　　　$x = 8$

　　$x + 6 = 14$　　　　　$x = 10$

　　$2x - 5 = 13$　　　　 $x = 7$

　　$4(x + 1) = 44$　　　 $x = 9$

11　Draw the region on a copy of the number line which satisfies both $x > 5$ and $x \leqslant 8$.

Now go back to the list of objectives at the start of this chapter.
How confident do you now feel about each of them?

8 Sequences

Go to www.collins.co.uk/edexcelpost16

Objectives

Before you start this chapter, mark how confident you feel about each of the statements below:	▶	▶▶	▶▶▶
I can recognise sequences of numbers, including odd and even numbers and Fibonacci sequences.			
I can write the term-to-term definition of a sequence in words.			
I can find a term in a sequence using position-to-term and term-to-term rules.			
I can draw the next term in a pattern sequence.			
I can find the nth term of a pattern sequence.			
I can find the nth term of a linear sequence.			
I can use the nth term of a quadratic sequence to generate terms.			
Stretch objectives			
I can continue a geometric progression and find the term-to-term rule.			

Check-in questions

- Complete these questions to assess how much you remember about each topic. Then mark your work using the answers at the back of the book.

- If you score well on all sections, you can go straight to the Revision checklist and Exam-style questions at the end of the chapter. If you don't score well, go to the chapter section indicated and work through the examples and practice questions there.

1 Give the next two terms in each sequence.

  a 1, 1, 2, 3, 5, 8, 13, ___, ___

  b 1, 4, 9, 16, 25, ___, ___

  c 1, 3, 6, 10, 15, ___, ___

  d 1, 3, 5, 7, 9, 11, ___, ___

  Go to 8.1

2 The cards show the nth terms of some sequences.

$2n$ $4n + 1$ $3n + 2$ $5n - 1$ $2 - n$

Match each sequence with the correct card.

a 5, 9, 13, 17…

b 1, 0, –1, –2…

c 2, 4, 6, 8, 10…

d 5, 8, 11, 14, 17…

e 4, 9, 14, 19…

Go to 8.2

3   **a**   Write down the next two terms in the sequence below.

     2   8   32   128   512

     b   Explain how to find the next number in the sequence. Go to 8.2

4   Here are the first four terms of an arithmetic sequence:

     5   7   9   11

     Write an expression in terms of n for the nth term of the sequence. Go to 8.2

8.1 Sequences

A **sequence** is a set of numbers that follow a particular rule. Any number that appears in a sequence is called a **term** of that sequence.

A **term-to-term rule** describes how to find the next number in a sequence using the previous term. For example, in the sequence 2, 4, 6, 8 …, you add 2 to the previous term to find the next term.

In an increasing sequence, the terms get larger. For example, 2, 7, 12, 17, 22 …

In a decreasing sequence, the terms get smaller. For example, 9, 6, 3, 0, –3, –6 …

Special sequences

Odd and even numbers

Odd numbers	1, 3, 5, 7, 9 …	nth term is $2n - 1$
Even numbers	2, 4, 6, 8, 10 …	nth term is $2n$

Square numbers

The sequence of square numbers is a **quadratic** sequence.

1	4	9	16	25	… nth term is n^2
$1^2 = 1 \times 1$	$2^2 = 2 \times 2$	$3^2 = 3 \times 3$	$4^2 = 4 \times 4$	$5^2 = 5 \times 5$	

See Chapter 1 for more on square numbers.

Fibonacci sequences

1, 1, 2, 3, 5, 8, 13 …   Add the previous two terms.

Other special sequences

You should also be able to recognise these sequences.

- 1, 8, 27, 64, 125...   cube numbers
- 1, 3, 6, 10, 15...   triangular numbers
- 2, 4, 8, 16, 32...   powers of 2
- 10, 100, 1000, 10 000, 100 000...   powers of 10

See Chapter 1 for more on cube numbers and triangular numbers.

Function machines and mapping

Function machines are useful when you need to find a relationship between two **variables**. For example, when numbers are fed into this machine they are first multiplied by 2 and then 1 is added.

Input (x) → $2x + 1$ →   Output (y)

If 1 is fed in, 3 comes out. $(1 \times 2 + 1 = 3)$

If 2 is fed in, 5 comes out. $(2 \times 2 + 1 = 5)$

Practice questions

1   The nth term of a sequence is $2n + 1$. Is this sequence odd numbers or even numbers? Give a reason for your choice.

2   The Fibonacci sequence is generated by adding together the previous two terms. Apply that rule to continue each of these sequences for two more terms.

  a 3, 4, 7 ...   **b** –4, –2 ...   **c** 0.5, 1 ...

3   Match each sequence with the correct name.

  1, 4, 9, 16 ...   cube numbers

  1, 3, 6, 10 ...   square numbers

  1, 8, 27, 64 ...   Fibonacci numbers

  1, 1, 2, 3, 5 ...   triangular numbers

4   In a function machine, each input is multiplied by 7 and then 3 is added. What is the output for each of these inputs?

  a 5   **b** 9   **c** 11   **d** –4

5   In a function machine, each input is multiplied by 3 and then 2 is subtracted. The output is 22. What was the input?

8.2 Finding the nth term of an arithmetic sequence

The nth term is the rule for a sequence and is often denoted by u_n. For example, the 8th term is represented as u_8.

In an arithmetic sequence, the terms increase (or decrease) by the same amount, called the common difference. For a **linear** or **arithmetic** sequence, the nth term is of the form: $u_n = an + b$.

The **position term** in a sequence is the general expression for the nth term which is related to the term's position in the sequence. It will give the value for the term in position n.

Example 1

Q Find the nth term of this sequence: 2, 6, 10, 14 …

A Look for the position term.

Position	1	2	3	4
Term	2	6	10	14

+4 +4 +4

The numbers are increasing in 4s. This means that the nth term is $4n$ + or − something.

When $n = 1$, $4n$ gives $4 \times 1 = 4$. Since the first term is 2, you need to subtract 2.

The rule is $4n - 2$.

Test this rule on the other terms:

$n = 2 \rightarrow (4 \times 2) - 2 = 6$

$n = 3 \rightarrow (4 \times 3) - 2 = 10$

It works on all of them, so the nth term is $4n - 2$.

The 20th term in the sequence would be $(4 \times 20) - 2 = 78$.

Exam tips The nth term of an arithmetic sequence is equal to common difference $\times n$ + zero term. (The zero term is the term that would be before the term in position 1.)

In a pattern sequence, the nth term can usually be found by looking at the practical context from which it arose.

For example:

Pattern 1   Pattern 2   Pattern 3

Number of squares	1	2	3	4	…
Number of matches	4	7	10	…	…

Example 2

Q a Draw pattern number 4 for the pattern sequence above.

b Write down the nth term, u_n.

c Alice has 51 matches. Is this enough matches to make 17 squares? Justify your answer.

A a

Pattern 4

b $u_n = 3n + 1$

This can be seen by looking at the structure of the shape: each additional square needs three more matches ($3n$) plus an extra one for the first square.

c No, Alice only has enough matches to make 16 squares.

$3n + 1 = 51$

$3n = 51 - 1$

$3n = 50$

$n = 16.6$

Example 3

Q The nth term of a sequence is $2n^2 + 1$. Chloe says that 101 is a number in the sequence.

Explain whether Chloe is correct.

A $2n^2 + 1 = 101$

$2n^2 = 101 - 1$

$2n^2 = 100$

$n^2 = 50$

Chloe is not correct. 101 is not in the sequence because 50 is not a square number.

Practice questions

1. The nth term of a sequence is given by $3n + 1$. Find the first four terms of the sequence.

2. Find the nth term of each sequence.

   a 2, 8, 14, 20 …

   b −1, 3, 7, 11, 15 …

   c 10, 18, 26, 34, 42 …

   d 101, 97, 93, 89 …

3. Copy these sequences and complete the missing terms.

   a 4, 8, ☐, 16, 20, ☐, 28

   b 23, ☐, 19, ☐, 15

   c 21, ☐, 63, ☐, 105

   d ☐, 8, 11, ☐, 17

4. A sequence has $2n^2 + 5$ as its nth term. Is 41 a number in the sequence? Explain how you know.

5. Find the value of the 50th term in the sequence 31, 39, 47, 55 …

REVISION CHECKLIST

- A sequence is a series of numbers that follow a particular rule.

- The nth term is the rule for a sequence, denoted by u_n, that enables you to work out any term in a sequence from its position in the sequence.

- The nth term of an arithmetic sequence is equal to common difference $\times n$ + zero term. (The zero term is the term that would be before the term in position 1.)

Exam-style questions

1. What type of numbers are these?

   1 8 27 64 125

even	odd	square	cube	triangular

2. Which two numbers are missing in this sequence?

   3, 7, ☐, 15, ☐, 23, 27 …

3 Work out the next two terms in this sequence.

20, 23, 26, 29 …

4 What are the first two terms in this sequence?

☐, ☐, 35, 32, 29, 26 …

5 What type of numbers are these?

2, 3, 5, 7, 11 …

6 What is the term-to-term rule for this sequence?

2, 6, 10, 14, 18 …

7 Find the 40th term of the sequence with nth term $5n - 6$.

8 The nth term of a sequence is $9 - 4n$.

Write down the first four terms.

9 Write down the nth term of this sequence.

9, 15, 21, 27, 33 …

10 The nth term of a sequence is $n^2 + 4$.

Work out the first term and the sixth term of the sequence.

11 The first five terms of a sequence are −2, 2, 6, 10 and 14.

a What kind of a sequence is this?

b Work out the 20th term.

12 Find the nth term for this sequence.

65, 61, 57, 53, 49 …

13 The nth term of a sequence is given by $2n - 6$.

Jose says that 56 is a number in the sequence. Is he correct? Show your working.

14 Write down the nth term of this sequence.

$\frac{1}{3}, \frac{3}{4}, \frac{5}{5}, \frac{7}{6}$ …

15 Here are two sequences.

3, 7, 11, 15, 19 …

50, 47, 44, 41 …

Will they have any terms in common? Show your working.

16 Here is a pattern of white and coloured dots.

a How many coloured dots will be in the 10th pattern?

b What is special about the sequence of numbers formed by the pattern of white dots?

Pattern 1   Pattern 2   Pattern 3

**Now go back to the list of objectives at the start of this chapter.
How confident do you now feel about each of them?**

9 Straight-line graphs

Check-in questions

- Complete these questions to assess how much you remember about each topic. Then mark your work using the answers at the back of the book.

- If you score well on all sections, you can go straight to the Revision checklist and Exam-style questions at the end of the chapter. If you don't score well, go to the chapter section indicated and work through the examples and practice questions there.

1 **a** Draw a set of axes which run from –10 to +10 in each direction. Plot these coordinates.

$$A = (5, -3) \qquad B = (-7, 9) \qquad C = (3, 8) \qquad D = (-9, -2)$$

b Write down the midpoints of the lines AB, AD and BD. *Go to 9.1*

2 **a** Copy and complete the table of values for $y = 2x + 3$.

x	–2	–1	0	1	2	3
y						

b Draw the graph of $y = 2x + 3$. *Go to 9.2*

3 Write down the gradient of the line $y = 4x - 3$. *Go to 9.3*

9.1 Coordinates

Coordinates are used to locate the position of a point. They are written in brackets and separated by a comma, e.g. (2, 4).

The first coordinate is the x coordinate and relates to the horizontal or x-axis. The second coordinate is the y-coordinate and relates to the vertical or y-axis.

When you read coordinates, read across first, then up (or down).

Example 1	**Q** Find the coordinates of A, B, C and D.

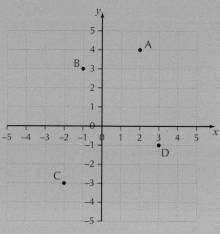

A A has coordinates (2, 4).

B has coordinates (–1, 3).

C has coordinates (–2, –3).

D has coordinates (3, –1).

Finding the midpoint of a line segment

You can find the coordinates of the midpoint of a line segment by working out the mean of the x coordinates and the y coordinates of the endpoints.

The midpoint of a line that joins the points $A(x_1, y_1)$ and

$B(x_2, y_2)$ is: $\left(\dfrac{(x_1+x_2)}{2}, \dfrac{(y_1+y_2)}{2} \right)$

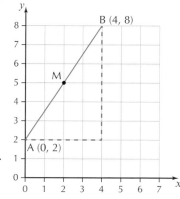

The midpoint, M, of the line AB in the diagram is $\left(\dfrac{(0+4)}{2}, \dfrac{(2+8)}{2} \right) = (2, 5)$.

Exam tips   It is helpful to learn the formula for finding the midpoint of a line segment.

Practice questions

1 What are the coordinates of Point A?

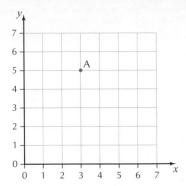

| (5, 3) | (5, 0) | (3, 0) | (3, 5) |

2 Draw a coordinate grid from –10 to +10 on both axes. Plot these coordinates.

A (5, 4) B (–3, 4) C (2, –8) D (–7, –2)

3 Draw a coordinate grid from –10 to +10 on both axes. Plot these coordinates.

A (5.5, 7) B (–0.5, 3) C (7.5, –8.1) D (–7.2, –2.4)

4 Using the points you plotted in Question 3, work out the midpoint of these line segments.

a AD **b** CD **c** BD **d** CB

5 A, B, C and D are line segments with the given endpoints. Three of the line segments have the same midpoint. Which line segment is the odd one out?

A (8, 1) and (4, 5) B (4, 1) and (7, 5) C (10, 3) and (2, 3) D (3, 0) and (9, 6)

9.2 Drawing straight-line graphs

You can draw a graph from its equation by working out the coordinates of some of the points.

To work out the coordinates for a graph, you can use either:

* a table (see examples)
* a function machine and mapping diagrams.

Graphs of the form $x = b$ and $y = a$

$y = a$ is a **horizontal** line, with every y coordinate equal to a.

$x = b$ is a **vertical** line, with every x coordinate equal to b.

Exam tips The equation of the x axis is $y = 0$.
The equation of the y axis is $x = 0$.

Example
2

Q a Draw the line $y = 3$. **b** Draw the line $x = 2$.

A

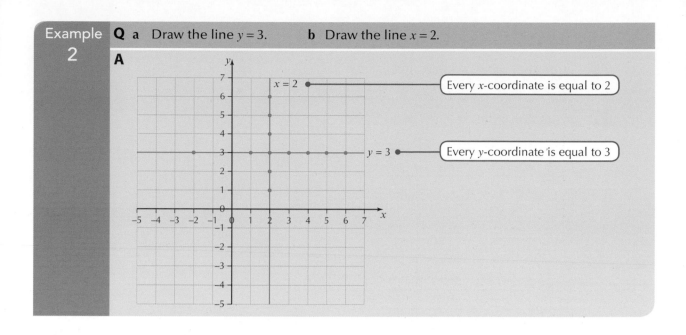

Every x-coordinate is equal to 2

Every y-coordinate is equal to 3

Graphs of the form $y = mx + c$

The general equation of a straight line is $y = mx + c$, where m is the **gradient** and c is the point where the line crosses the y-axis, also known as the y-**intercept** $(0, c)$.

To draw the graph of $y = 3x - 4$:

- Draw a table of values, substitute the x values into the equation $y = 3x - 4$ (e.g. $x = 2$, $y = 3 \times 2 - 4 = 2$) and record the corresponding y values.

x	−1	0	2	4
y	−7	−4	2	8

- Plot the points $(-1, -7)$, $(0, -4)$, $(2, 2)$ and $(4, 8)$ on a pair of axes. Remember across first, then up/down.

- Join them with a straight line and add labels.

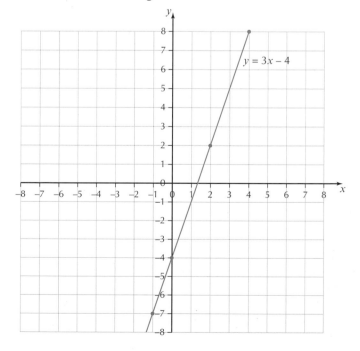

$y = 3x - 4$

Exam tips

If no table is given in the question, draw up a table of x values and work out the corresponding y values. Choose a minimum of three points. It can be useful to work out the value of y when $x = 0$ and the value of x when $y = 0$ as two of the points.

Practice questions

1   Draw these lines on the same pair of axes.

     $y = 3$ $y = 5$ $x = 7$ $x = 9$

2   Write down the equation of each line.

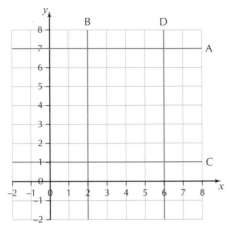

3   Copy and complete each table of values.

a

$y = x + 3$						
x	−2	−1	0	1	2	3
y						

b

$y = 2x - 3$						
x	−2	−1	0	1	2	3
y						

c

$y = -\frac{1}{2}x + 2$						
x	−2	−1	0	1	2	3
y						

4   Draw the graph of $y = 2x - 9$. What is the value of x when $y = 5$?

5   Draw the graph of $y = -x - 6$. Using your graph, find the value of x when $y = 3$.

9.3 Calculating the gradient of a straight-line graph

Before you calculate the gradient of a graph, make sure you have looked carefully at the scales on both axes. They may not be the same.

You can use this formula to calculate the gradient of a graph:

$$\text{gradient} = \frac{\text{change in } y}{\text{change in } x}$$

For the straight line in the diagram: gradient $= \dfrac{6}{4}$

$$= \dfrac{3}{2}$$

$$= 1.5$$

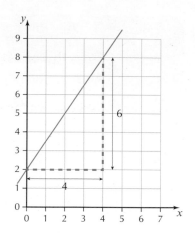

Remember the equation of a straight line is of the form $y = mx + c$, where m is the gradient and c is the **intercept** on the y-axis.

The equation of the line is $y = 1.5x + 2$.

Parallel lines have the same gradients. $y = 2x - 3$, $y = 2x + 1$ and $y = 2x - 1$ are three parallel lines with a gradient of 2.

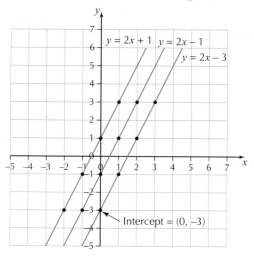

Intercept $= (0, -3)$

If the gradient (m) is positive, the line slopes upwards (from bottom left to top right). If m is negative, the line slopes downwards (from top left to bottom right).

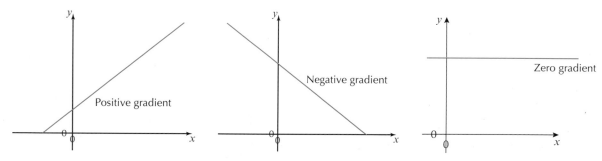

Positive gradient

Negative gradient

Zero gradient

The greater the gradient, the steeper the line.

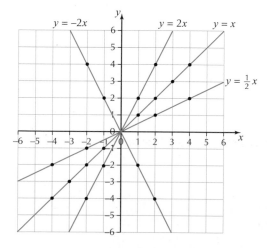

Example 3

Q Write down the gradient and y-intercept for each straight-line graph.

   a $y = 4x - 3$ **b** $y = 6 - 2x$ **c** $2y + 10 = 4x$

A **a** Gradient $= 4$, Intercept $= (0, -3)$

   b Gradient $= -2$, Intercept $= (0, 6)$

   c Gradient $= 2$, Intercept $= (0, -5)$

> This equation needs to be rearranged into the form $y = mx + c$:
> $$2y = 4x - 10$$
> $$y = 2x - 5$$

Exam tips Check the scales of the axes carefully. One square does not always represent one unit. The scales on the horizontal and vertical axes do not need to be the same.

Practice questions

1 Work out the gradient of each line.

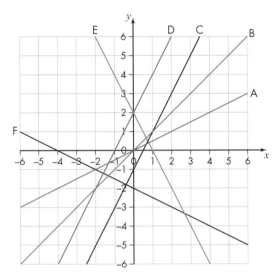

2 Write down the gradient of each line.

   a $y = 3x + 6$ **b** $y = x - 3$ **c** $y = \frac{1}{2}x - 7$ **d** $y = -x + 4$

3 Write down all the pairs of parallel lines.

   A: $y = 3x - 4$ B: $y = -4x + 6$ C: $y = 4x - 3$

   D: $y = 4x - 6$ E: $y = 3x - 3$ F: $y = -4x - 4$

> **Hint**
>
> Parallel lines have the same gradient.

4 Write down the y-intercept of each line.

   a $y = 3x + 6$ **b** $y = x - 3$ **c** $y = \frac{1}{2}x - 7$ **d** $y = -x + 4$

5 Write down the gradient and y-intercept of the line with equation $2y = 4x - 6$.

REVISION CHECKLIST

- The general equation for a straight line is $y = mx + c$, where m is the gradient and c is the y-intercept.

- Gradient $= \dfrac{\text{change in } y}{\text{change in } x}$.

- To work out the midpoint of a line segment, find the mean of the x coordinates and the mean of the y coordinates.

- Parallel lines have the same gradient.

Exam-style questions

1 What are the coordinates of Point A?

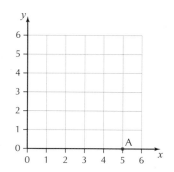

(5, 3)	(5, 0)	(3, 0)	(3, 5)

2 Draw the lines with equations $y = -2$ and $x = 3$ on a coordinate grid.
Write down the coordinates of the point of intersection.

3 In which region of the grid does each point fall?

    a (3, 4)   **b** (−2, 1)   **c** (−6, −7)   **d** (2, −0.5)

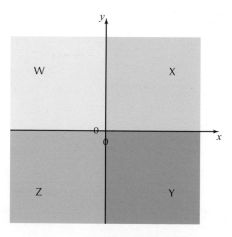

4 Write down the equation of a line that is parallel to the line $y = -6$.

5 Match each equation with the appropriate description.

$y = 5$	horizontal
$x = -2$	positive gradient
$y = 3x - 4$	vertical
$2y = -2x - 1$	negative gradient

6 Work out the coordinates of the midpoint of the line that joins A(3, 9) and B(5, 13).

7 Work out the coordinates of the midpoint of the line that joins A(–2, 8) and B(7, 9).

8 The points with coordinates (2, 4), (4, 2) and (6, 4) form three vertices of a square. What are the coordinates of the fourth vertex?

9 Which of these equations represents a straight line? Write down all that apply.

A: $y = 5$ B: $y = 3x + 2$ C: $y = x^2 + 3$ D: $y = 3^x$ E: $x = -1$

10 What is the gradient of the line with equation $2y = -4x + 3$?

–4	2	3	–2	1	1.5

11 Which of these lines has the steepest gradient?

A: $y = -x + 3$ B: $y = 2x - 5$ C: $y = 6x + 4$ D: $y = -10x + 1$ E: $y = -0.5x + 12$

12 Write down the gradient and y intercept for the line with equation $5y = 6x + 10$.

13 Work out the gradient of the line AB.

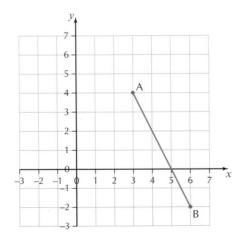

14 A line, L, is parallel to EF but passes through (0, 7). Work out the equation of L.

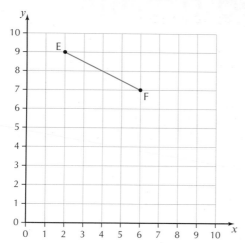

15 Write the equation of a line with the steepest possible gradient.

16 Which of these lines are parallel?

A: $y = 3x + 6$ B: $3y = x + 6$ C: $3y = 9x + 12$ D: $y = 6x + 3$

17 A line has gradient 2 and intercept –5. What is the equation of the line?

**Now go back to the list of objectives at the start of this chapter.
How confident do you now feel about each of them?**

10 Curved graphs

Objectives

Before you start this chapter, mark how confident you feel about each of the statements below:

	▶	▶▶	▶▶▶
I can generate and plot the points of a quadratic graph.			
I can identify the line of symmetry of a quadratic graph.			
I can find approximate solutions to quadratic equations using a graph.			
I can identify and interpret roots, intercepts and turning points of quadratic graphs.			
I can recognise and sketch simple cubic functions.			
I can recognise and sketch graphs of the reciprocal function $y = \dfrac{1}{x}$.			

Check-in questions

- Complete these questions to assess how much you remember about each topic. Then mark your work using the answers at the back of the book.
- If you score well on all sections, you can go straight to the Revision checklist and Exam-style questions at the end of the chapter. If you don't score well, go to the chapter section indicated and work through the examples and practice questions there.

1. Which of these equations represents a ∪ shaped curve?

   A: $y = -5x^2 + 9$ B: $y = 5x^2 - 9$ C: $y = 5x - 9$ D: $y = -5x + 9$

   Go to 10.1

2. Draw the graph of $y = x^2 + x - 2$, using x values from −3 to +3.

   Go to 10.1

3. Draw a coordinate grid from −10 to +10 on both axes. Draw the graph of $y = x^2 + 3x - 4$. Use your graph to solve $x^2 + 3x - 4 = 5$.

   Go to 10.2

4 Match each graph below with the correct equation.

Go to 10.3

$$y = x^3 - 5 \qquad y = 2 - x^2 \qquad y = 4x + 2 \qquad y = \frac{3}{x}$$

Graph A

Graph B

Graph C

Graph D

10.1 Quadratic graphs

The equations of **quadratic graphs** are of the form $y = ax^2 + bx + c$ where $a \neq 0$. Quadratic graphs have an x^2 term as the highest power of x.

They will be $\cup$ shaped if the **coefficient** of x^2 is positive, and $\cap$ shaped if the coefficient of x^2 is negative.

| Example 1 | **Q** Draw the graph of $y = x^2 - 2x - 6$. |

A • Draw up a table of values and complete it by substituting the values of x into the equation, for example for $x = 1$, $y = 1^2 - 2 \times 1 - 6 = -7$, so coordinates are $(1, -7)$.

x	−2	−1	0	1	2	3	4
y	2	−3	−6	−7	−6	−3	2

• Draw a coordinate grid using values of x from −2 to 4 and corresponding values of y on graph paper and plot the points.

• Join the points with a smooth curve.

• Label the curve.

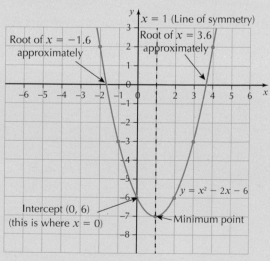

• Every quadratic graph has a **turning point**. If $a > 0$, the turning point will be the **minimum** point. If $a < 0$, the turning point will be the **maximum** point. For the example above, the turning point is at $(1, -7)$ and is a minimum point. The minimum value of y is −7.

- The **line of symmetry** occurs at $x = -\frac{b}{2a}$. For the example above, $b = -2$ and $a = 1$, so $x = -\frac{-2}{2 \times 1} = 1$.

  The equation of the line of symmetry is $x = 1$.

You will need to be able to identify both of these.

> **Exam tips** Choose a scale that makes it easy to plot the points.
> It is a good idea to use brackets when you need to square a negative number, for example $(-3)^2$.

Practice questions

1. Write down the coordinates of the minimum point on this graph.

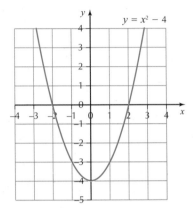

2. Copy and complete the table of values for $y = x^2 + x - 3$.

x	−2	−1	0	1	2	3	4
y							

3. Write down the equation of the line of symmetry for this graph.

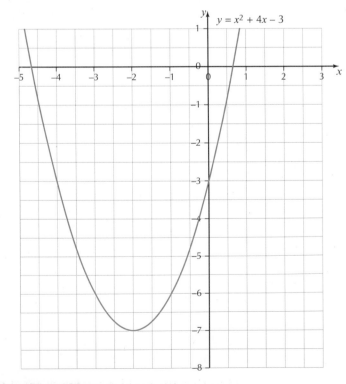

4 Draw these graphs.

    a $y = x^2 + 2x - 4$ **b** $y = x^2 + 3x - 3$ **c** $y = x^2 - 3x - 2$ **d** $y = -x^2 - 2x + 8$

5 Draw these graphs. For each graph, write down:
    i the coordinates of the turning point
    ii the equation of line of symmetry.

    a $y = x^2 + 3x - 4$ **b** $y = -x^2 - 3x + 4$ **c** $y = 2x^2 + x - 1$ **d** $y = -2x^2 + 3x - 4$

10.2 Using graphs to solve quadratic equations

You can use a graph to solve one or more equations. The important points are the points of intersection either with the x-axis or with each other.

Example 2

Q The graph shows $y = x^2 - 8x + 7$.

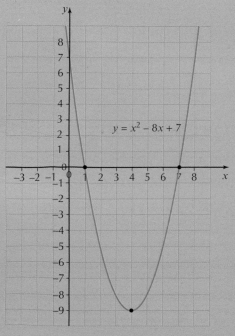

    a Use the graph to solve $x^2 - 8x + 7 = 0$.

    b Use the graph to solve $x^2 - 8x + 7 = -8$.

A   **a** The solutions to the equation $x^2 - 8x + 7 = 0$ are the points where the quadratic curve crosses the x-axis, so the solutions are $x = 1$ and $x = 7$. These solutions are called the **roots** of the equation.

(continued)

b To find the value(s) of x when $y = -8$, draw a line at $y = -8$ and read the x-values of the points of intersection with the graph: $x = 3$ and $x = 5$.

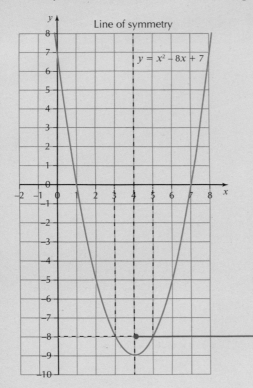

Read across from $y = -8$ to find the x-coordinates of the points of intersection of the graph and the line $y = -8$.

Exam tips To answer these questions correctly, draw your curve as accurately as possible. Show how you have found values using dashed lines.

Practice questions

1 Use this graph to find the roots of the equation $x^2 - 2x - 3 = 0$.

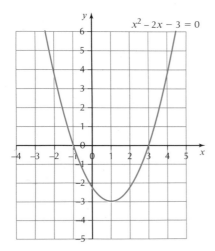

2 Draw the graph of $y = x^2 + x - 2$ and use it to find the roots of the equation $x^2 + x - 2 = 0$.

3 Draw a graph and use it to solve the equation $x^2 + x - 12 = 0$.

4 Draw the graph of $y = x^2 + 2x - 4$ and the line $y = 5$. Explain how you would use these two graphs to solve $x^2 + 2x - 9 = 0$ graphically.

5 Draw the graph of $y = x^2 + 2x + 4$. Explain why the equation $x^2 + 2x + 4 = 0$ has no roots.

10.3 Graph shapes

You need to be able to recognise certain types of graphs from their shapes. These are illustrated in the diagrams below.

$y = mx + c$

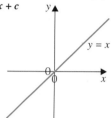

$y = ax^2 + bx + c$ $\quad y = x^2$

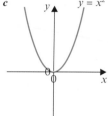

$y = \frac{k}{x}$

(Reciprocal functions) $\quad y = \frac{1}{x}$

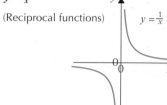

$y = x^3$

(Cubic functions) $\quad y = x^3$

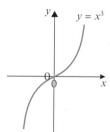

$y = ax^3 + bx^2 + cx + d$ $\quad y = 2x^3 + 5x^2 - 2x - 3$

(Cubic functions)

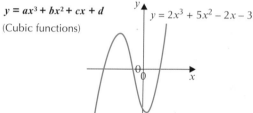

Example 3

Q Match each graph with its equation.

$y = x^2 - 4 \qquad y = 5 - 2x \qquad y = x^3 \qquad y = 3x - 1$

Graph A

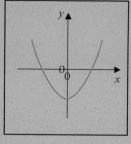

Graph B

Graph C

Graph D

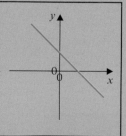
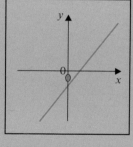

A Graph A is $y = x^3$

Graph B is $y = 3x - 1$

Graph C is $y = x^2 - 4$

Graph D is $y = 5 - 2x$

Example
4

Q **a** Draw and complete a table of values for $y = x^3 - 1$.

 b Choose a suitable scale and then draw the graph of $y = x^3 - 1$.

 c From the graph, find the approximate value of x when $y = 15$.

A **a**

x	−3	−2	−1	0	1	2	3
y	−28	−9	−2	−1	0	7	26

b

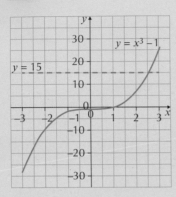

c $x = 2.5$

Practice questions

1 Match each graph with the correct equation.

$y = 2x^2 + 1$	$y = \dfrac{1}{2}x^2 + 2x + 1$	$y = -x^3 + 1$	$y = \dfrac{1}{x}$

Graph A

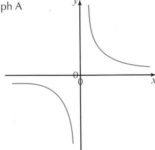

Graph B

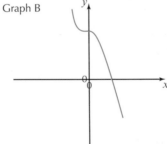

Graph C

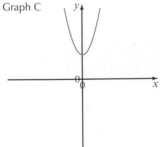

Graph D

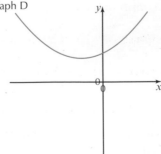

2 Which graph will have steeper sides, $y = 3x^2 + 1$ or $y = \dfrac{1}{4}x^2 + 2x + 1$? Explain your choice.

3 Draw the graph of $y = \dfrac{2}{x-2}$.

4 **a** Draw the graph of $y = 1 - x^3$.

    b Use the graph to find an approximate solution when $y = -5$.

5 **a** Draw the graphs of $y = 2x + 6$ and $y = \dfrac{1}{x+1}$.

    b Use the graphs to find approximate solutions to $2x + 6 = \dfrac{1}{x+1}$.

REVISION CHECKLIST

- Make sure you know the different graph shapes.
- Quadratic graphs have an x^2 term as the highest power of x. They are $\cup$ shaped or $\cap$ shaped.
- The coordinates of the points where the graph crosses the x-axis are the roots of the equation.
- Cubic graphs have an x^3 term as the highest power of x.
- The general form of the reciprocal function is $y = \dfrac{k}{x}$.

Exam-style questions

1 Copy and complete the table of values for $y = x^2 + 3x - 4$.

x	−2	−1	0	1	2
y	−6			0	

2 **a** Copy and complete the table of values for $y = x^2 + x + 1$.

x	−2	−1	0	1	2
y					

    b Draw the graph of $y = x^2 + x + 1$ for values of x from −3 to +3.

3 Draw the graph of $y = 2x^2 - 1$ for values of x from −2 to +2.

4 Draw the graph of $y = x^2 + 3x - 2$ for values of x from −4 to +1.

5 Sketch the graph of $y = x^2 + 3x + 6$. Explain why it has no roots.

6 Sketch the graph of $y = x^3 + 2$. How many roots does this graph have?

7 How many roots are there for the equation $-x^2 - 2x - 3 = 0$?

0	1	2	3	4

8 The graph of $y = x^2 + 2x - 24$ has roots at $x = -6$ and $x = 4$. Write down the equation of the line of symmetry.

9 Jamal says that every quadratic graph will cut the x-axis twice. Sketch a quadratic graph that shows this is not true.

10 **a** Copy and complete the table of values for $y = x^3 - 3$.

x	−2	−1	0	1	2	3
y	−11			−2		

b Draw the graph of $y = x^3 - 3$.

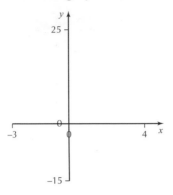

c How many roots does the equation $x^3 - 3 = 0$ have?

11 **a** Copy and complete the table of values for $y = \dfrac{2}{x}$.

x	−3	−2	−1	−0.5	−0.4	−0.3	−0.2	0.2	0.3	0.4	0.5	1	2	3
y	$-\dfrac{2}{3}$			−4			−10							

b Draw the graph of $y = \dfrac{2}{x}$.

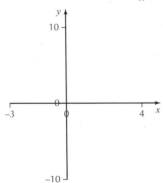

c Explain why there is no value for y when $x = 0$.

12 State whether each graph is linear, quadratic, cubic, reciprocal or none of these.

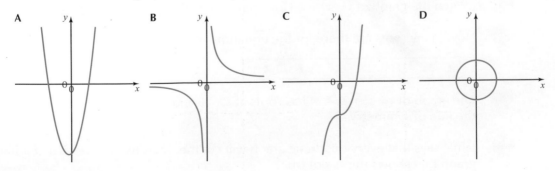

A   B   C   D

13 State whether each graph is linear, quadratic, cubic, reciprocal or none of these.

A

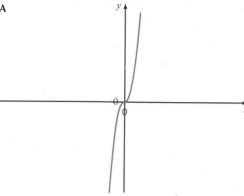

B

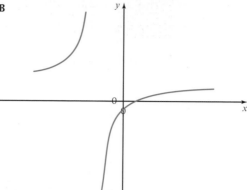

C

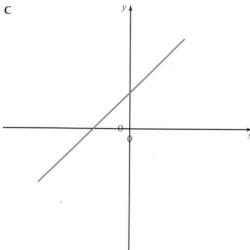

D

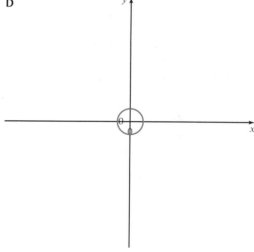

14 Sketch a cubic graph with three roots.

15 Which of these show a negative cubic graph?

A

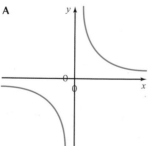

B

C

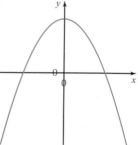

D

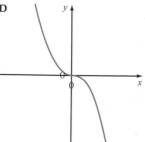

E

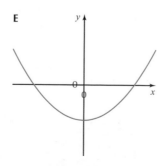

16 **a** Factorise $x^2 - 4x - 5$.

   b Write down the roots of the equation $x^2 - 4x - 5 = 0$.

   c Write down the coordinates of the y-intercept of the graph with equation $y = x^2 - 4x - 5$.

   d Sketch the graph of $y = x^2 - 4x - 5$.

17 Work out the equation of the line of symmetry on the graph of $y = 2x^2 + x - 6$.

18 **a** Draw the graph of $y = x^2 - 5x - 6$.

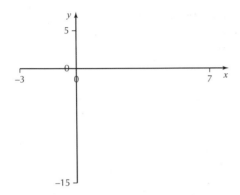

   b Write down the equation of the line you need to draw on the same axes so that you can use the graphs to solve $x^2 - 5x + 3 = 0$.

Now go back to the list of objectives at the start of this chapter.

How confident do you now feel about each of them?

11 Interpreting graphs

Objectives

Before you start this chapter, mark how confident you feel about each of the statements below:

	▶	▶▶	▶▶▶
I can read values from straight-line graphs representing real-life situations.			
I can draw real-life graphs such as ones to use to calculate rates of discount or charges to be made.			
I can draw and interpret conversion graphs.			
I can draw and interpret distance–time graphs.			
I can interpret speed–time graphs.			

Check-in questions

- Complete these questions to assess how much you remember about each topic. Then mark your work using the answers at the back of the book.
- If you score well on all sections, you can go straight to the Revision checklist and Exam-style questions at the end of the chapter. If you don't score well, go to the chapter section indicated and work through the examples and practice questions there.

1 Use the graph to answer the questions. Go to 11.1

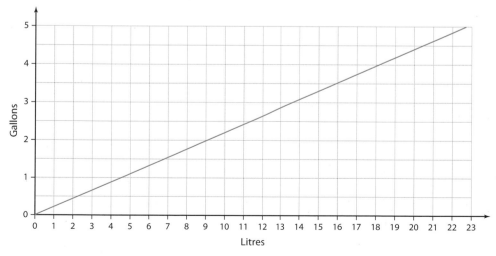

a How many litres are in 3 gallons?

b How many gallons are in 20 litres?

2 The diagram shows the temperature in Ally's flat. The temperature is recorded and plotted every 2 hours.

Go to 11.2

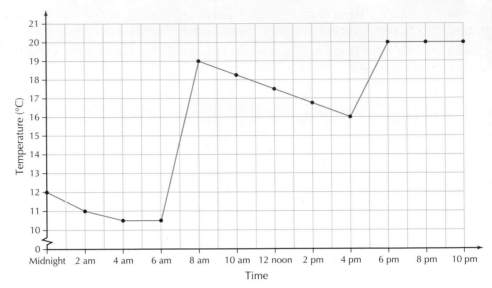

a The heating system comes on automatically in the morning to warm the flat. At what time do you think it is it set to come on?

b What is the highest temperature recorded in the flat?

3 The distance–time graph shows the car journeys of two people.

Go to 11.3

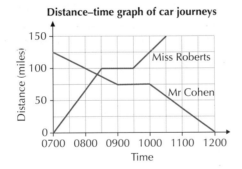

Distance–time graph of car journeys

Answer the questions using the graph.

a Work out Miss Roberts' speed in the first 100 miles.

b For how long was Mr Cohen stationary?

c Work out Mr Cohen's speed between 1000 and 1200.

d At approximately what time did Miss Roberts and Mr Cohen pass each other?

e Work out Miss Roberts' speed between 0930 and 1030.

11.1 Conversion graphs

Conversion graphs are used to change amounts in one unit of measurement to another unit, for example: litres to pints, kilometres to miles, pounds to dollars.

Because these units are in direct proportion to each other, you can draw a straight line conversion graph using only one piece of information.

For example, if £1 = $1.50 you can draw this graph:

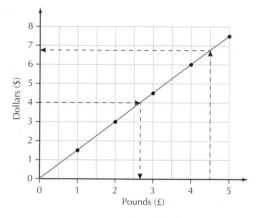

To change an amount in dollars to pounds, read across to the line and then read down: for example, $4 ≈ £2.70.

To change an amount in pounds to dollars, read up to the line then read across: for example, £4.50 ≈ $6.80.

The conversion rate is always changing, so a graph like the one above will never be accurate for long. However, it is a useful tool to give an approximate idea of price differences.

Exam tips Draw lines on the diagram to demonstrate to the examiner how you have read your answers from the graph. Remember to check you have used the scale correctly.

Practice questions

1 This conversion graph can be used to change between pounds (£) and dollars ($)

   a Change $5 to pounds.

   b Change £2 to dollars.

   c Explain how to work out the conversion rate used in this graph.

   d Explain how you could use the graph to convert £50 into dollars.

   e In Britain, a bottle of moisturiser costs £40. The same product in America costs $46. In which country is it cheaper and by how much?

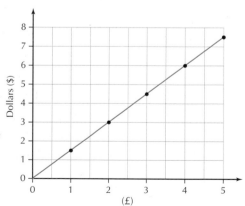

2 One mile is approximately 1.6 km.

   a Draw a conversion graph for distances from 0 to 10 miles.

   b Using you graph, convert 5 miles into kilometres.

   c Using your graph, convert 12 km into miles.

   d Explain three different ways of converting 20 miles into kilometres. Check each way gives you the same answer.

   e John and Casey work in the same office. John lives 23 miles away and Casey lives 38 km away. Use your graph to work out which one of them lives further from the office.

11.2 Real-life graphs

A **linear graph** can be used to represent real-life situations.

For some questions, you will need to interpret the gradient of a straight-line graph. A straight-line graph has a constant gradient. It tells you how much the variable on the vertical axis increases or decreases when you increase the amount on the horizontal axis by one unit.

| Example 1 | **Q** Charlotte and some friends are travelling to a music festival. They decide to hire a car. |

The graph shows the charges made by the car-hire firm.

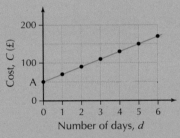

Exam tips

Look carefully at the scales on the axes. Do not just count the squares on the graph, since the scale on each axis may be different.

   a What do you think point A represents?

   b By using the gradient of the line, work out how much was charged per day for the hire of the car.

   c Write down a formula that connects the cost (C) of the car hire and the number of days (d).

   d Charlotte and her friends hire the car for 3 days. What is the total cost of the car hire?

A **a** Point A represents the initial £50 charge for hiring the car. This is paid regardless of the hire period.

   b Gradient = 20, so the charge per day is £20.

   c $C = 20d + 50$ ———• Notice that this equation is in the form of $y = mx + c$.

   d $C = 20d + 50$

   $= 20 \times 3 + 50$

   $= £110$ ———• You could also read this from the graph. Read up from 3 on the horizontal axis, until it meets the line, then read across.

The gradient of the line shows how fast the quantity is changing. A straight line implies a constant rate of change. A horizontal line means there is no change.

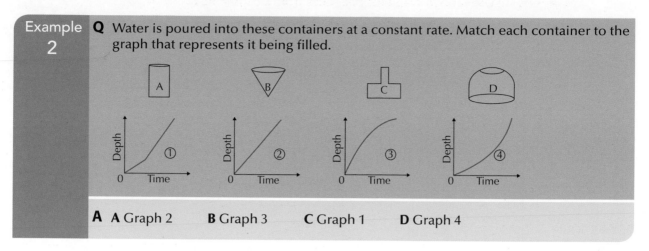

Example 2

Q Water is poured into these containers at a constant rate. Match each container to the graph that represents it being filled.

A **A** Graph 2   **B** Graph 3   **C** Graph 1   **D** Graph 4

Practice questions

1 Simon pours water into these containers at a steady rate.

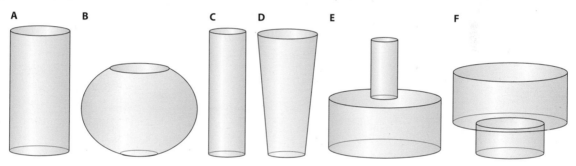

He measures the depth of water in the containers over time and then draws these graphs on the same scales.

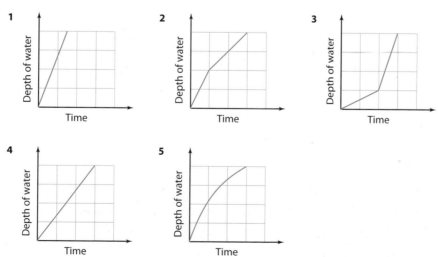

a Match the graphs to the containers. One container has no graph.

b Sketch a graph for the remaining container.

2 The diagram shows the temperature in Ally's flat. The temperature is recorded and plotted every 2 hours.

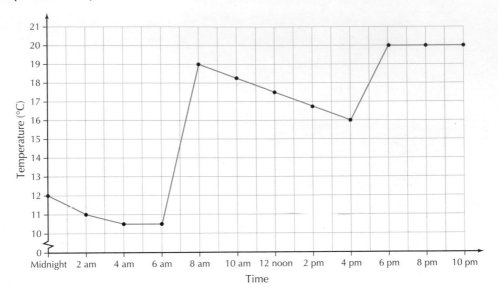

James says that when he came in from work, the flat was at 22 °C so he turned down the heating. Could this be true? You must explain your answer.

3 The graph shows the amount of water (in litres) in Dom's bath.

Dom starts to run the bath at 0 minutes. He stops running the bath when the doorbell rings.

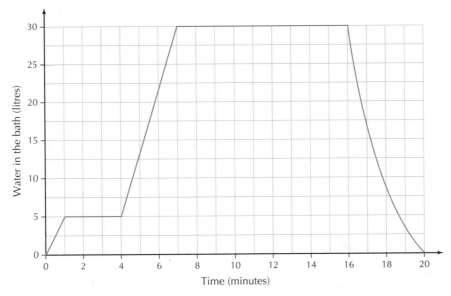

a After how many minutes does the doorbell ring?

b How long after the doorbell rings does he turn the taps back on?

c After how many minutes (from 0) is the bath filled?

d How long does the bath take to empty?

4 The graph shows the cost of parking (£) in a city-centre car park for different lengths of time (hours).

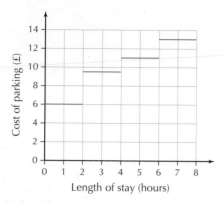

a What is the charge for staying 3 hours?

b Tilly had planned to stay for 2 hours but ended up staying for 3 hours. How much more did this cost her?

c Gavin has £8. What is the maximum length of time he can stay in the car park?

5 The graph shows the mouse population over time in a particular field. Describe what has happened to the mouse population over time.

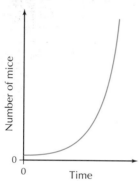

11.3 Distance–time and velocity–time graphs

Distance–time graphs

Distance–time graphs are often known as travel graphs. Time is shown on the horizontal axis; distance is shown on the vertical axis.

The **gradient** of the distance–time graph represents the average **speed** of the object over that time interval.

Always check that you understand the scales on a distance–time graph.

Example
3

Q Mr Smith travels from Leeds to his office 80 miles away.

The distance–time graph shows his journey.

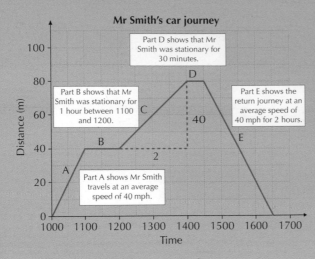

Mr Smith's car journey

Part D shows that Mr Smith was stationary for 30 minutes.

Part B shows that Mr Smith was stationary for 1 hour between 1100 and 1200.

Part E shows the return journey at an average speed of 40 mph for 2 hours.

Part A shows Mr Smith travels at an average speed of 40 mph.

Work out his average speed for part C of the journey.

A On the horizontal axis, one square represents 30 minutes (half an hour). On the vertical axis, one square represents 10 miles.

$$\text{Average speed} = \frac{\text{distance travelled}}{\text{time taken}}$$

In part C of the graph, average speed $= \dfrac{40}{2}$

$= 20$ mph

Exam tips When drawing return journeys on a distance–time graph, you cannot go back in time, so the line does not go back to the start.

Velocity–time graphs

Velocity means speed in a certain direction. A **velocity–time** graph shows the rate of change of velocity with time.

You can use a **velocity–time** graph to work out the distance travelled by an object or the **acceleration** or **deceleration** of an object.

- The area between the graph and x-axis represents the distance travelled.

- The gradient represents the acceleration at a given time.

  - A positive gradient means the velocity is increasing.
  - A negative gradient means the velocity is decreasing.
  - A horizontal line means the velocity is constant.

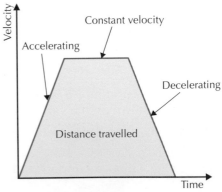

Exam tips Remember to include the units with your answers.

Practice questions

1 Farouk drove from his house to the airport. He set off at 10:00 a.m. and stopped on the way for a break. The distance–time graph shows his journey.

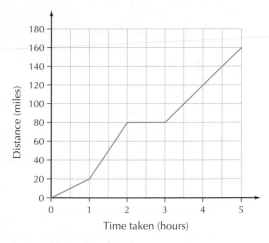

a At what time did Farouk: **i** stop for his break   **ii** arrive at the airport?

b What is the distance between Farouk's home and the airport?

c What was his average speed for: **i** the first hour   **ii** the last part of his journey?

2   **a** Calculate the average speed during each stage of the journey shown in the graph.

   i A to B   **ii** B to C   **iii** C to D   **iv** D to E

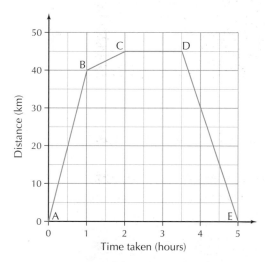

b How can you tell, just by looking at the graph, in which part of the journey the speed was greatest?

3   Work out the average speed of each journey.

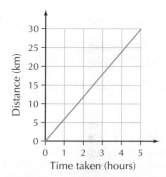

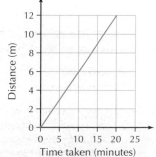

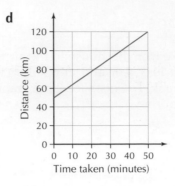

c

Distance (km) vs Time taken (hours)

d

Distance (km) vs Time taken (minutes)

4 The graph shows a race between Rob and Darren.

Describe the race.

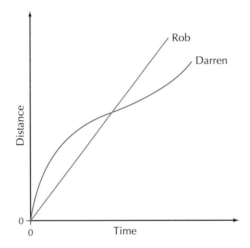

5 Deepak cycles to work via the shops, a total distance of 18 km.

It takes him 30 minutes to cycle the 8 km to the shops. He stops for 10 minutes at the shops and then continues his journey to work. He arrives at work one hour and fifteen minutes after he left home.

Draw a distance–time graph to show his journey. Assume that he cycled at a constant speed for the whole journey.

REVISION CHECKLIST

- The gradient of a distance–time graph represents the speed.

  $$\text{Average speed} = \frac{\text{distance travelled}}{\text{time taken}}$$

- You can find the acceleration or deceleration of an object using a velocity–time graph. This is represented by the gradient of the graph at a given time.

- The area under a velocity–time graph represents the distance travelled by the object.

- Check the scales on graphs carefully by noting the value of each square.

Exam-style questions

1 Jez drives from home to a training session. This is the distance–time graph for his complete journey.

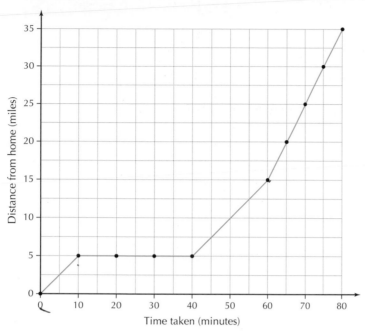

a For how long was he stopped in traffic?

b How far from his home was the training session?

c What was his average speed for the whole journey?

2 This graph shows a conversion between pounds (lb) and kilograms (kg).

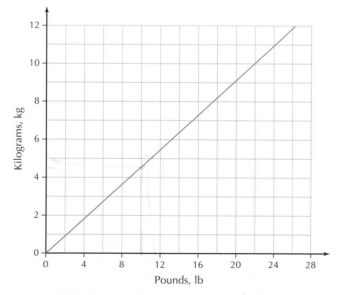

a Convert 5 kg into pounds.

b Convert 10 lb into kilograms.

3   The travel graph shows the distance a plane has travelled from an airport over time. After how long has the plane travelled 1650 miles?

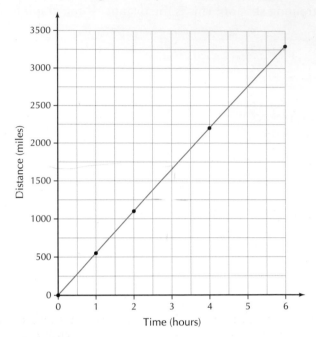

4   This conversion graph shows the relationship between distances in miles and kilometres.

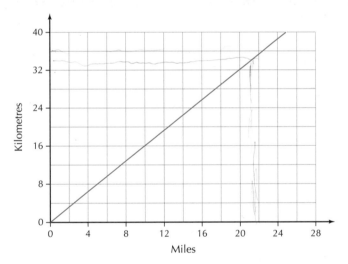

Lucia drives 35 km visiting clients. On the same day, Toni drives 20 miles.

Which of them has driven further? You must explain your answer.

5 The graph shows the depth of water in a harbour, measured every hour after high tide which occurred at midnight.

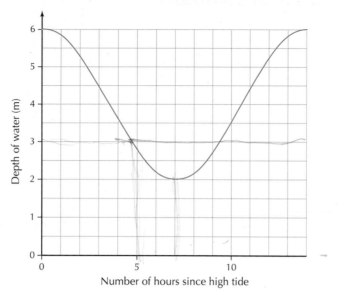

Number of hours since high tide

a After how many hours is low tide?

b Scott's boat can only enter the harbour when there is at least 3 m of water. He left to go fishing at midnight and wants to return at 5 a.m. Is this possible? You must explain your answer.

6 The conversion graph shows the relationship between temperatures in degrees Celsius and degrees Fahrenheit.

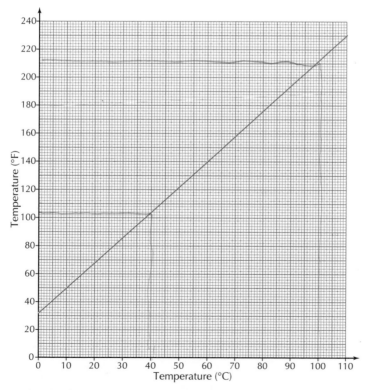

Temperature (°C)

a Water boils at 100 °C. What is this in degrees Fahrenheit?

b A healthy human has a temperature of 37 °C. Jermaine has a temperature of 102 °F. Is his temperature too high, too low or just right? You must explain your answer.

7 Work out the average speed of the journey shown in this graph. Remember to state the units of your answer.

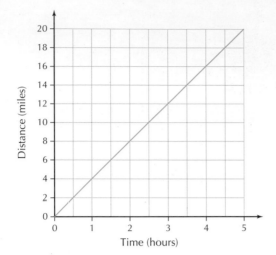

8 The graph shows the charges made for various distances by a taxi firm.

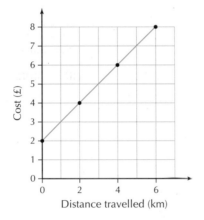

a How much would the company charge for a journey of 6 km?

b i Work out the gradient of this graph.

  ii What does the gradient represent?

9 This conversion graph shows the relationship between distances in miles and kilometres.

Carlton can claim expenses for his journey at a rate of 45p per mile. He has travelled 100 km. Work out how much Carlton can claim for the journey.

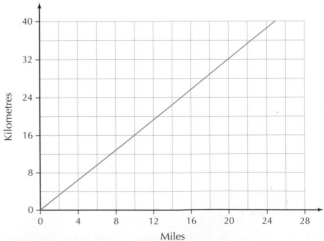

10 George needs a new wheelchair. He is choosing from the three models shown below.

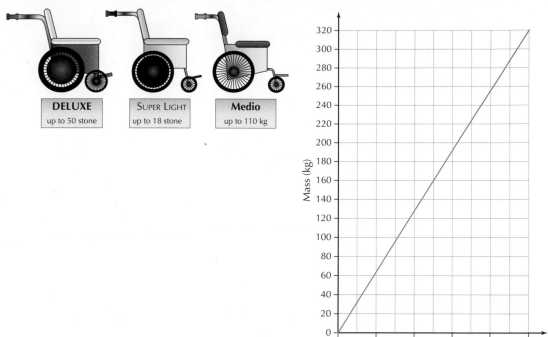

He would like to buy the Medio wheelchair, but is not sure if it will be suitable as he weighs 18 stone. Should he buy the Medio? Justify your answer.

11 Betty is training for a race. This is the distance–time graph for her training session.

a Between which period of time does she run fastest? Explain your answer.

b For how long does she stop to rest?

c What was her average speed for the first 2 minutes?

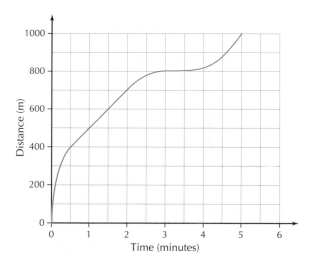

12 Matthew and Eddie take part in a race. This is the distance–time graph.

   a Who was faster at the start of the race?

   b After how long did the person in second place move to the front?

   c Who won the race?

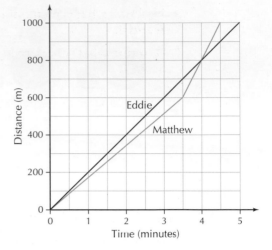

13 The graph shows LongTax's charges for taxi journeys of different distances.

A second company, CabLong, has no flat charge and charges £1.50 per kilometre.

   a Copy the grid and draw a graph to show the charges for CabLong.

   b For what distance is the cost the same for both firms?

   c Niles wants to travel 5 km. Which firm should he use? You must explain your answer.

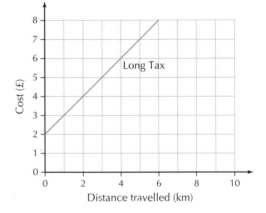

14 Julia makes and delivers wedding cakes. She has a cake to deliver to a customer that lives 30 miles away. She drives there at a constant speed and the journey takes her 45 minutes. She has a cup of tea with the client which takes 20 minutes and then returns home, again driving at a constant speed. She arrives back at home 100 minutes after she first left.

   a Draw a distance–time graph on a copy of the grid to represent her journey.

   b Work out her average speed on the return journey.

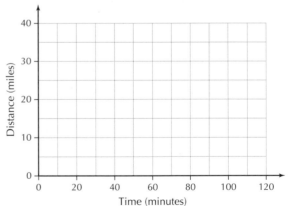

15 Here is a velocity–time graph.

   a Describe what is happening between A and B.

   b Describe what is happening between B and C.

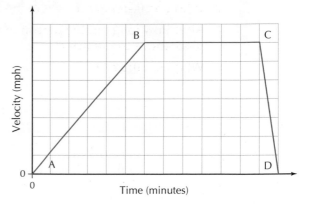

16 Archie fires an arrow. The velocity–time graph shows the journey of the arrow.

   a What is the velocity of the arrow when it leaves the bow?

   b For how long does the arrow fly?

   c Describe the journey of the arrow.

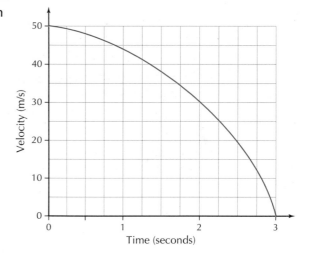

Now go back to the list of objectives at the start of this chapter. How confident do you now feel about each of them?

12 Percentages

Check-in questions

- Complete these questions to assess how much you remember about each topic. Then mark your work using the answers at the back of the book.
- If you score well on all sections, you can go straight to the Revision checklist and Exam-style questions at the end of the chapter. If you don't score well, go to the chapter section indicated and work through the examples and practice questions there.

1 Work out these quantities.　　　　　　　　　　　Go to 12.1

    a 12% of 50 kg　　**b** 30% of £2000　　**c** 5% of £60　　**d** 35% of 720 g

2 Work out these. You may use a calculator.　　　Go to 12.1

    a 16% of 2.5 kg　　**b** 56% of 9250 g　　**c** 83% of £6 000 000

3 Reduce £225 by 20%.　　　　　　　　　　　　Go to 12.2

4 The full price of a jumper is £65. In a sale, it is reduced by 35%. How much does the jumper cost in the sale?　　Go to 12.2

5 Express 32 as a percentage of 40.　　　　　　　Go to 12.3

6 Charlotte earns £96 per week babysitting. She spends £18 of this on train fares. What percentage of her earnings is this?

Go to 12.3

7 Alfie buys a car for £8500. He sells it two years later for £4105. Work out the percentage loss. Give your answer to three significant figures.

Go to 12.4

8 Emmie paid £85 000 for her flat in 2014. The flat increased in value by 12% in 2015 and by 28% in 2016. How much was her flat worth at the end of 2016?

Go to 12.5

9 Ryan is saving for a car. He invests £3500 in a bond that earns 2.3% per year compound interest. How much interest will he earn in two years?

Go to 12.6

10 Each item listed below includes value added tax (VAT) at 20%. Work out the cost of each item before VAT is added.

Go to 12.7

a A pair of shoes: £69   b A coat: £152.40

c A suit: £285   d A television: £525

12.1 Percentage of a quantity

- **Percentages** are parts of 100.
- This is the percentage sign: %
- The examples show different methods for working out percentages of quantities. You will need to be able to use a calculator and a non-calculator method.

Example 1	**Q** Work out 30% of 80 kg.

A Method 1: Calculator

$\frac{30}{100} \times 80 = 24$ kg

Key in: 3 0 ÷ 1 0 0 × 8 0 =

or: 0 . 3 × 8 0 = •——— $30\% = \frac{30}{100} = 0.3$, which is known as the **multiplier**.

or: 8 0 × 3 0 % =

Method 2: Non calculator

10% of 80 kg = 80 ÷ 10 •——— Find 10% by dividing by 10.

= 8 kg

3 × 8 = 24 kg •——— Multiply by 3 to find 30%.

Exam tips Percentage questions set on a calculator paper are easier to work out using a calculator.

Practice questions

1 Work out these quantities.

   a 30% of 70   **b** 60% of 290   **c** 20% of 3490   **d** 80% of 640

2 Work out these quantities.

   a 5% of 260   **b** 15% of 290   **c** 35% of 150   **d** 95% of 6480

3 Use a calculator to work out these quantities.

   a 7% of 2500   **b** 11% of £3800   **c** 23% of 4900 kg **d** 47% of 456 m

4 Use a calculator to work out these quantities.

   a 16.5% of 280 g **b** 43.8% of 265 km   **c** 21.7% of $48   **d** 20.4% of £900

5 Use a calculator to work out these quantities. Give your answers to one decimal place.

   a 123% of 49 m **b** 201% of 26 kg   **c** 155% of £30   **d** 186.4% of 94 kg

12.2 Increasing and decreasing quantities by a percentage

You need to find be able to find the value of quantities after a percentage increase or decrease. This is a useful skill for daily life.

Example 2	**Q** A tablet costs £165. In a sale, it is reduced by 15%. Work out the cost of the tablet in the sale.

A Method 1

$$15\% \text{ of } £165 = \frac{15}{100} \times 165$$

$$= £24.75$$

The tablet costs £165 − £24.75 = £140.25 in the sale.

Method 2: Using a multiplier

100 − 15 = 85   ←   (If the original price is 100%, the sale price is 100 − 15 = 85%.)

The tablet costs 0.85 × 165 = £140.25 in the sale.   ←   (0.85 is the multiplier – the price is going down.)

Exam tips   The original price will always be 100%. If the price is reduced, such as in a sale, the new price will be less than 100%. If the price is increased, such as by adding VAT, the new price will be more than 100%.

Example 3

Q Louisa works out the cost of her gas bill. At the start of a three-month period, the gas meter reading was 12 447 units. At the end of the three-month period, the gas meter reading was 12 721.

Each unit of gas used costs 47p. VAT is charged at 5%. Work out the total cost of Louisa's gas bill.

A 12 721 − 12 447 = 274 units used

274 × 47p = £128.78 cost of gas without VAT

VAT at 5%, so price plus VAT = 105%

Multiplier 1 + 0.05 = 1.05 ●————— | 1.05 is the multiplier – the price is going up. |

Total gas bill is 1.05 × 128.78 = £135.22 ●———— | Round to the nearest penny. |

Exam tips Remember to subtract the initial reading from the final reading to find the number of units used in questions like Example 3. Always check your answer makes sense. For example, a gas bill of £13 522 is unlikely.

Practice questions

1 Increase each quantity by the percentage shown.

   a £150 by 13%   **b** 300 g by 40%   **c** 250 m by 37%   **d** 652 kg by 64%

2 Decrease each quantity by the percentage shown.

   a £52 by 37%   **b** 46 kg by 76%   **c** 307 km by 23%   **d** 4408 mm by 44%

3 Find the multiplier for each percentage change.

   a 15% decrease   **b** 43% increase   **c** 76% decrease   **d** 88% increase

4 Tal wants to buy a jacket in the sale. He has £70 in vouchers. Before the sale, the jacket cost £95. In the sale, the jacket is reduced by 22%. Can Tal buy the jacket using only his vouchers? Show your working.

5 Katya wants to buy a new laptop. The model she wants costs £540 in Techstore and £445 + 20% VAT in Comp Solutions. In which shop is it cheaper? Show your working.

12.3 One quantity as a percentage of another

When you are asked to compare different quantities, it is useful to write each quantity as a percentage.

To express one quantity as a percentage of another, divide the first quantity by the second quantity and multiply by 100%.

Example 4

Q There is 3.6 g of fat in a pint of semi-skimmed milk. 2.2 g of the fat is saturated. What percentage of the fat is saturated?

A $\frac{2.2}{3.6}$ •———— First write the amount of saturated as a fraction of the total amount of fat.

$\frac{2.2}{3.6} \times 100\% = 61.1\%$ •———— Convert the fraction to a percentage by multiplying by 100.

Example 5

Q Matthew got 46 out of 75 in his science test and 65% in his maths test. In which test did he get a better score?

A Work out his score in the science test as a percentage.

$\frac{46}{75} \times 100\% = 61.\dot{3}\%$

To work this out on the calculator, key in: 4   6   ÷   7   5   ×   1   0   0   =

Matthew got a better score in the maths test because 65% > 61.3%.

Practice questions

1 Write the first number as a percentage of the second number.

  a 3, 60   **b** 5, 25   **c** 9, 20   **d** 13, 40

2 Write the first number as a percentage of the second number. Give your answers to one decimal place.

  a 8, 40   **b** 15, 35   **c** 19, 70   **d** 131, 400

3 Rosa sits two maths exams. She scores 62 out of 75 in the first exam and 80 out of 105 in the second exam. Did she score better in the first or second exam? Show your working.

4 Maggie is looking at book reviews on two different websites. On the "BooksRUs" website, the book is rated by readers as 4.5 out of 5. On the "ReadALot" website, the same book is rated as 18.6 out of 20. Which website rated the book higher?

5 The Christmas special hot chocolate contains 11.3 g of sugar in a 226 g serving. What percentage of the drink is sugar?

12.4 Percentage change

Sometimes you are asked to calculate a percentage profit, percentage loss, percentage increase or percentage decrease. This is referred to as a percentage change and can be found as follows:

$$\text{percentage change} = \frac{\text{change}}{\text{original}} \times 100$$

Example 6

Q Rich bought a flat for £185 000.

He sold it three years later for £242 000.

What is his percentage profit?

A His actual profit is £242 000 − £185 000 = £57 000

His percentage profit is $\frac{57\,000}{185\,000} \times 100 = 30.8\%$ (3 s.f.)

Example 7

Q Jackie bought a car for £12 500 and sold it two years later for £7250.
Work out her percentage loss.

A Her actual loss is £12 500 − £7250 = £5250

Her percentage loss is $\frac{5250}{12\,500} \times 100 = 42\%$

Example 8

Q Varia sold a necklace on an internet auction site for £225. She originally paid £65 for the necklace. What was her percentage profit, to the nearest one percent?

A Her actual profit was £225 − £65 = £160

Percentage profit was $\frac{160}{65} \times 100 = 246\%$

Exam tips It is possible to have a percentage change that is more than 100% if the actual change is more than the original amount.

Practice questions

1 Work out the percentage change.

	Start value	End value	Percentage change
a	150	180	
b	30	42	
c	320	280	
d	450	150	

2 Work out the percentage change.

	Start value	End value	Percentage change
a	1150	1800	
b	3000	4250	
c	320 000	280 000	
d	3450	150	

3 John bought a house for £60 000 in 1997.
He sold the house for £195 000 in 2016.
What was his percentage profit?

4 Skylar paid £12 500 for her car in 2014.
She sold the car in 2016 for £6500.
What was her percentage loss?

5 In January, Tara's puppy weighed 5 kg. By August, the dog weighed 12 kg. What was the dog's percentage weight gain?

12.5 Repeated percentage change

A quantity can increase or decrease in value each year by a percentage. If the percentage change is repeated over successive years, the original value changes at the end of each year. Example 9 shows two methods of calculating repeated percentage change. The second method uses a **multiplier** and is usually quicker.

Example 9	**Q** Alex paid £12 500 for a car. Each year it depreciated in value by 15%. What was the car worth after three years?

A Method 1

Depreciation of 15% is a value of 100% − 15% = 85%

Value at end of year 1: $\dfrac{85}{100} \times £12\,500 = £10\,625$

Value at end of year 2: $\dfrac{85}{100} \times £10\,625 = £9031.25$ ● — £10 625 depreciates in value by 15%.

Value at end of year 3: $\dfrac{85}{100} \times £9031.25 = £7676.56$ ● — £9301.25 depreciates in value by 15%.

The car is worth £7676.56 after three years.

Method 2: Using a multiplier

Finding 85% of the value of the car is the same as multiplying by 0.85

Value at end of year 1: $0.85 \times £12\,500 = £10\,625$

Value at end of year 2: $0.85 \times £10\,625 = £9031.25$

Value at end of year 3: $0.85 \times £9031.25 = £7676.56$

This is the same as $(0.85)^3 \times £12\,500 = £7676.56$

The car is worth £7676.56 after three years.

Exam tips A reduction of 15% each year for three years is not a reduction of 45%!

Practice questions

1 Work out the amount of money in each account.

	Original amount	Annual interest rate	Number of years invested	Final amount
a	500	3%	3	
b	1350	2%	4	
c	8420	1.6%	5	
d	10 000	4.3%	6	

2 A kitten gains 10% of its body mass each week for the first eight weeks. Work out the mass of a kitten after eight weeks if its mass at birth was 100 g.

3 In 2009, experts estimated the UK hedgehog population to be one million hedgehogs. Hedgehogs are now on the endangered species list as their numbers are falling by about 5% per year. Estimate the size of the UK hedgehog population in 2019.

4 Bill buys some hedges to create a screen in his garden. They are 1 m tall when he buys them and the hedges are said to grow by 20% each year. He wants an 8 m screen. How long will the hedges take to reach this height?

5 Chelsea invests £500 in shares. She makes an average profit of 7% per year. Estimate her total profit after 4 years.

12.6 Interest and tax

Simple interest

With **simple interest**, interest is paid only on the original amount invested. The interest is the same amount each year. It is a straightforward percentage change calculation.

Example 10	**Q** £3200 is invested for 4 years at 3.2% per annum.
	a How much simple interest is paid in total after 4 years?
	b How much is in the account after 4 years?
	A a $\frac{3.2}{100} \times 3200 = £102.40$ per year
	$4 \times £102.40 = £409.60$ paid for 4 years
	b $£3200 + £409.60 = £3609.60$

Compound interest

When a bank or account offers **compound interest**, interest is paid on the interest already earned as well as on the original amount invested. It is a repeated percentage change.

Example 11	**Q** Becky has £3200 in her savings account. She receives compound interest at 3.2% per annum.
	How much will she have in her account after 4 years?
	A 100% + 3.2% = 103.2%. This is equivalent to a multiplier of 1.032.
	At end of year 1: $1.032 \times £3200 = £3302.40$
	At end of year 2: $1.032 \times £3302.40 = £3408.08$
	At end of year 3: $1.032 \times £3408.08 = £3517.14$
	At end of year 4: $1.032 \times £3517.14 = £3629.68$
	A quicker way is to multiply £3200 by $(1.032)^4$
	Number of years
	£3200 $\times$ $(1.032)^4 = £3629.68$
	Original Multiplier
	She will have £3629.68 after 4 years.

Example
12

Q Shamil invests £3000 in each of two bank accounts. The terms of the bank accounts are shown below.

Shamil says that he will earn the same amount of interest from both bank accounts in 2 years.

Decide whether Shamil is correct. Show full working to justify your answer.

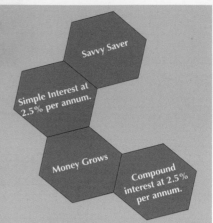

Savvy Saver

Simple Interest at 2.5% per annum.

Money Grows

Compound interest at 2.5% per annum.

A Annual interest from Savvy Saver: $\dfrac{2.5}{100} \times 3000 = £75$

Total interest from Savvy Saver in 2 years: $£75 \times 2 = £150$

Total amount in Money Grows after 2 years: $1.025^2 \times 3000 = £3151.88$

Total interest earned in 2 years: $£3151.88 - £3000 = £151.88$

Shamil is not correct – he would earn £1.88 more with the Money Grows investment.

> **Exam tips** Using a multiplier enables you to answer the question efficiently.

National Insurance (NI)

National Insurance is usually deducted as a percentage of a wage or salary.

Example
13

Q Manisha earns £1575 a month. National Insurance at 11% is deducted. How much National Insurance must she pay?

A 11% of £1575 $= 0.11 \times 1575$

$= £173.25$

Income tax

A percentage of a wage or salary is deducted for **income tax**. Before tax is calculated, a personal allowance (that is not taxable) is subtracted.

The standard personal allowance is £11 500. Income tax is then charged at different rates.

The table shows the tax rates for each band in 2016–2017 if you have a standard personal allowance of £11 500.

Band	Taxable income	Tax rate
Personal allowance	Up to £11 500	0%
Basic rate	£11 501 to £45 000	20%
Higher rate	£45 001 to £150 000	40%
Additional rate	over £150 000	45%

Example
14

Q Will earns £190 per week. The first £62 is not taxable but the remainder is taxed at 20%. How much income tax does he pay each week?

A Taxable income: £190 − £62 = £128

20% of taxable income: 0.2 × 128 = £25.60

Will pays £25.60 tax per week.

Practice questions

1 Work out the difference between the simple interest and compound interest paid at a rate of 4.5% on an investment of £600 for 3 years.

2 Lior earns £1200 per month. He pays National Insurance of 11%. How much is this?

3 Charis earns £175 per week. The first £62 is not taxable. She pays 20% tax on her remaining income. How much tax does she pay?

4 The CEO of a company earns £260 000 per year. Her personal allowance is £7800. The table shows the rates of tax payable. Calculate her annual tax bill.

Rate of tax	Tax band
Basic rate 20%	Up to and including £37 400
Higher rate 40%	Over £37 400

5 Pami and Mandeep each pay the same rate of income tax. Pami has a salary of £32 000 and a personal allowance of £7400. Mandeep has a salary of £37 500 and a personal allowance of £5400. Which of them pays more tax?

12.7 Reverse percentage problems

In **reverse** percentage problems, you are asked to find the value of the original quantity given the amount after a percentage increase or decrease.

Start by working out what percentage of the original amount the quantity you are given represents.

Example 15

Q The price of a television is reduced by 15% in a sale to £352.75.

What was the original price?

A £352.75 represents 100% − 15% = 85% of the original price (x).

The multiplier for 85% is 0.85.

$0.85 \times x = £352.75$

$x = £352.75 \div 0.85$

$= £415$

The original price of the television was £415.

Check:

original price → × 0.85 → new price

original price ← ÷ 0.85 ← new price

Exam tips Think carefully about whether the amount you are trying to find is larger or smaller than the amount you are given. Check that your answer sounds sensible. An original price should always be more than a sale price.

Q A mobile phone bill is £169.20 including tax at 20%. What is the cost of the bill without the tax?

A £169.20 represents 100% + 20% = 120% of the bill without tax (x).

The multiplier for 120% is 1.2.

$1.2 \times x = £169.20$

$\qquad x = £169.20 \div 1.2$

$\qquad = £141$

The bill without tax is £141.

Check:

original bill $\rightarrow \times 1.2 \rightarrow$ new bill

original bill $\leftarrow \div 1.2 \leftarrow$ new bill

Q A tablet costs £60 after a 15% reduction. Joseph says that the original price of the table was £70.59.

Is Joseph correct? Show your working.

A £60 represents 100% − 15% = 85% of the original price (x).

The multiplier for 85% is 0.85.

$0.85 \times x = £60.00$

$\qquad x = £60.00 \div 0.85$

$\qquad x = £70.59$

Joseph is correct.

Check:

original price $\rightarrow \times 0.85 \rightarrow$ new price

original price $\leftarrow \div 0.85 \leftarrow$ new price

Practice questions

1. A washing machine costs £144 in a 20% off sale. Calculate the original cost of the washing machine.

2. Dhiaan is given a pay rise of 5%. Her new monthly salary is £1260. What was her monthly salary before the pay rise?

3. Carlton buys a new car. He gets a discount of 12% on the original price and pays £13 200. What was the original price of the car?

4. Felicity buys a pair of boots in the sale. She pays £95 and works out that she saves approximately 27%. What was the cost of the boots before the sale to the nearest pound?

5. Anita pays £59 for an annual on-demand TV service. She joined the service during their sale period and saved 25% off the normal cost. What is the normal cost of the service?

REVISION CHECKLIST

- Percentages are parts of 100. To find a percentage of a quantity, write the percentage as a fraction (denominator 100) and multiply by the quantity.

- It may be useful to remember common equivalences when working with fractions, decimals or percentages.

- To express one quantity as a percentage of another, divide the first quantity by the second quantity and multiply by 100. The result can be greater than 100%.

- Use a multiplier to work out a repeated percentage change.

- Simple interest is paid on an investment each year. The same amount is paid each year.

- Compound interest is paid on the interest already earned as well as on the original investment.

- In a reverse percentage problem, you have to find the original amount from the amount after a given percentage change. You can use a multiplier to work out reverse percentage problems.

- Remember to check that your answers seem sensible.

Exam-style questions

1. Which is greater, 41% of 50 or 51% of 40? You must explain your answer.

2. Aneke plants 50 tomato seeds. 30% of them do not grow. How many seeds do not grow?

3. Brianna has 250 albums. 48 of them are Jazz. What percentage is this?

4. Kev makes and sells bath fizzers. He wants to sell them for the same price as last year so he makes them 10% smaller. Last year, the mass of the large size was 48 g. What is the mass of the large size this year?

5. Last year 2000 people attended the village festival. Half of them bought a burger. This year Josie expects a 17% increase in the number of people attending and the same percentage to buy burgers. How many burgers should she order for this year?

6. Norah leaves the house at 7:30 a.m. and returns after work at 7 p.m. What percentage of her day is spent out of the house?

7. A cereal bar with a mass of 37 g contains 12 g of sugar. What percentage of the cereal bar is sugar?

8. Write 48 as a percentage of 250.

9. China won 26 gold medals in the 2016 Olympic games. Seven of these were in diving. What percentage of their medals were for diving?

10. Matt bought a flat for £180 000 and sold it for £210 000.

    Rik bought a flat for £120 000 and sold it for £150 000.

    Which of them made the greater percentage profit?

11 Alison bought 80 aubergines to sell at a market. She paid 60p per aubergine and put them up for sale for £1.10 each. She sold 30 of the aubergines at this price, and the remaining aubergines at half price. What was her overall percentage profit or loss on the sale of the aubergines?

12 Louis paid £1.70 each for some shares. When he sold them, they were only worth £1.24 per share. What was his percentage loss?

13 Adam is a mechanic. He buys a car for £200 and pays £100 to fit a new clutch. He sells the car for £950. What is his overall percentage profit?

14 Ally is advised to increase her daily step count gradually. On Monday, she takes 4000 steps and plans to increase this by 5% each day.

   a How many steps should she do on Tuesday?

   b After how many days will she be doing more than 6000 steps?

15 Elise invests £400 at a rate of 6%, paid every 3 months.

   a How much will she have after one year?

   b How long will it take for her to have earned over £100 in interest?

16 The value of a car, originally worth £14 000, falls by 17% each year. How much will the car be worth after 5 years?

17 Ernest has a salary of £32 000 per year. He has a personal tax allowance of £10 800 and pays tax at rate of 20% on the rest of his income. How much tax will he pay each month?

18 There are 1.5 million people living in Kent. This is approximately 2.3% of the population of the UK. What is the approximate population of the UK?

19 Ozzy and Bailey each want to buy the same model of phone. Ozzy pays £156, including VAT at 20%. Bailey pays £125 plus VAT at 20%. Which of them paid less?

20 A box of chocolates is on special offer. The mass of the chocolates in the box is increased by 20%. The mass of chocolates in the special offer box is 240 g. What was the mass of the chocolates in the box before the special offer?

Now go back to the list of objectives at the start of this chapter.
How confident do you now feel about each of them?

13 Ratio and proportion

Objectives

Before you start this chapter, mark how confident you feel about each of the statements below:

	▶	▶▶	▶▶▶
I can write ratios in their simplest form.			
I can write ratios in the form $1 : m$ or $m : 1$.			
I can write a multiplicative relationship between two quantities as a ratio or fraction, e.g. if a and b are in the ratio $4 : 7$, then a is $\frac{4}{7} b$.			
I can compare ratios.			
I can share a quantity in a given ratio.			
I can solve proportion problems using the unitary method.			
I can solve a ratio problem in context, including identifying better buys.			
I can work with fractions in ratio problems.			
I can solve problems using direct and inverse proportion.			

Stretch objectives	Go to www.collins.co.uk/edexcelpost16		
I can set up and use equations to solve problems involving direct and inverse proportion.			

Check-in questions

- Complete these questions to assess how much you remember about each topic. Then mark your work using the answers at the back of the book.

- If you score well on all sections, you can go straight to the Revision checklist and Exam-style questions at the end of the chapter. If you don't score well, go to the chapter section indicated and work through the examples and practice questions there.

1 Which one of these ratios is the odd one out?

| $300 : 400$ | $76 : 100$ | $21 : 28$ | $39 : 52$ | $60 : 80$ |

Go to 13.1 ▶

2 Divide £160 in the ratio $1 : 2 : 5$.

Go to 13.2 ▶

3 I pay £6.72 for 4 ringbinders. How much will 21 ringbinders cost?

Go to 13.3 ▶

4 A shop sells two sizes of Rice Pops.
Which pack is better value for money?

Go to 13.3

600 g

£2.40

250 g

£1.10

5 It took six builders four days to lay a patio. How long would
it take eight builders working at the same rate?

Go to 13.4

13.1 Simplifying ratios

A **proportion** is part of the whole. It is usually written as a fraction or percentage.

A **ratio** is used to compare two or more related quantities. We can read the ratio symbol ':' as 'compared to', so 16 boys compared to 20 girls can be written as 16 : 20 in ratio form. You can simplify ratios by dividing both parts by the highest common factor.

Example 1	**Q** Simplify the ratio 16 : 20.
	A 16 : 20 = 4 : 5 ⟵ (Divide both sides by 4.)

Exam tips Ratios can be cancelled down like fractions into their simplest form, e.g. $\frac{16}{20} = \frac{4}{5}$.

Example 2	**Q** Simplify the ratio 21 : 28.
	A 21 : 28 = 3 : 4 ⟵ (Divide both sides by 7.)

Example 3	**Q** Express the ratio 5 : 2 in the ratio m : 1.
	A $5 : 2 = \frac{5}{2} : \frac{2}{2}$ ⟵ (Divide both parts by 2.)
	$= 2.5 : 1$

<table>
<tr><td>Example
4</td><td>Q Two quantities, p and q, are in the ratio $3:5$. Express:
a p as a fraction of q b q as a fraction of p.</td></tr>
<tr><td></td><td>A $p:q = 3:5$
    a $p = \dfrac{3}{5}q$ b $q = \dfrac{5}{3}p$</td></tr>
</table>

Exam tips   It can be helpful to write both ratios as fractions first, e.g. $\dfrac{p}{q} = \dfrac{3}{5}$, and then multiply by the denominator, in this case $p = \dfrac{3}{5}q$.

Exam tips   Check that the quantities are expressed in the same units before you write and simplify a ratio. In **similar** shapes, corresponding sides are in the same ratio. (See Chapter 15 for more on similarity.)

Practice questions

1   Simplify these ratios.

    a $6:9$ **b** $12:20$ **c** $32:14$ **d** $18:81$

2   Write these ratios in the form $1:m$.

    a $14:42$ **b** $3:12$ **c** $15:30$ **d** $500:75$

3   Write these ratios in the form $m:1$.

    a $14:42$ **b** $3:12$ **c** $15:30$ **d** $500:75$

4   Which ratio is the odd one out?

| $25:35$ | $40:56$ | $255:357$ | $15:20$ |

5   Simplify the ratio 5 minutes : 1 hour.

13.2 Sharing a quantity in a given ratio

To share an amount in a given ratio, add up the individual parts and then divide the amount by this number to find the value of one part.

<table>
<tr><td>Example
5</td><td>Q £155 is divided in the ratio of $2:3$ between Daisy and Tom. How much does each receive?</td></tr>
<tr><td></td><td>A $2 + 3 = 5$ parts   •————(Add up the total parts.)
     5 parts = £155
       1 part = £155 ÷ 5   •————(Work out what one part is worth.)
           = £31
     So Daisy gets $2 \times$ £31 = £62 and Tom gets $3 \times$ £31 = £93
     Check: £62 + £93 = £155 ✓</td></tr>
</table>

Example 6

Q A business makes a profit of £32 000. The profit is divided between the directors in the ratio 3 : 2 : 5. How much do they each receive?

A 3 + 2 + 5 = 10 parts

10 parts = £32 000

1 part = 32 000 ÷ 10

= £3200

So the directors get:
3 × 3200 = £9600

2 × 3200 = £6400

5 × 3200 = £16 000

Check: £9600 + £6400 + £16 000 = £32 000 ✓

Example 7

Q Kuschal and Jack take part in a sponsored charity bike ride. They share the money collected between their charities in the ratio 3 : 7.

Jack receives £336 more for his charity than Kuschal.

How much money was there in total?

A Jack gets 4 more parts than Kuschal.

4 parts = £336

1 part = 336 ÷ 4

= £84

Total number of parts is 3 + 7 = 10

10 parts = £840

There was £840 in total.

Exam tips Always check your answers by comparing the total amount with the sum of the individual parts.

Practice questions

1 Share £48 in these ratios.

   a 1 : 2   **b** 5 : 7   **c** 1 : 3   **d** 5 : 1

2 Louise and Will share £50 in the ratio 7 : 3.

How much will each of them receive?

3 Tamsin orders some soft toys to re-stock her shop. She orders 100 soft toys – dolls, teddies and animals – in the ratio 2 : 3 : 5 respectively. How many more animals than dolls does she order?

4 A festival has 4000 visitors. The ratio of male to female visitors was 3 : 2.

How many females visited the festival?

5 Ed and Ted share the running costs of a car in the ratio 4 : 3 with Ed paying the larger share. They receive a repair bill for £234.56. How much is Ted's share?

13.3 Problem solving

Two quantities are in **direct proportion** when both quantities increase at the same rate, for example the number of hours worked per week and the weekly wage.

Example 8	**Q** Samuel went on holiday to Spain. He changed £350 into euros.

Q Samuel went on holiday to Spain. He changed £350 into euros.

The exchange rate was £1 = €1.36.

a How many euros did Samuel receive?

b When Samuel returned home, he still had €112. The exchange rate was then £1 = €1.24. How many pounds did Samuel receive when he changed the remaining money?

A **a** £350 = 350 × €1.36

= €476

b €112 = 112 ÷ 1.24

= £90.32

Example 9

Q Jessica buys a pair of jeans in Britain for £52.

When she goes to America, she sees an identical pair of jeans for $63.

The exchange rate is £1 = $1.49.

In which country are the jeans cheaper, and by how much?

A You can either convert the price in pounds to dollars:

£52 × $1.49 = $77.48

$77.48 − $63 = $14.48

or convert the price in dollars into pounds:

$63 ÷ 1.49 = £42.28

£52 − £42.28 = £9.72

The jeans are cheaper in America by £9.72 (or by $14.48).

Exam tips It is very important that you include the units of currency in your answer.

Best buys

There is usually more than one way to answer these questions, but working out the unit cost will help you decide which is better value for money.

Example 10

Q The same brand of breakfast cereal is sold in two different-sized packets. Which packet represents better value for money?

Work out the cost per gram for each box of cereal.

(continued)

A 125 g costs £0.89, so $\frac{89}{125}$ = 0.712p per gram.

500 g costs £2.10, so $\frac{210}{500}$ = 0.42p per gram.

The larger box is better value for money because it costs less per gram.

Example 11

Q Toothpaste is sold in three different-sized tubes.

50 ml costs £1.24

75 ml costs £1.96

100 ml costs £2.42

Which size is the best value for money? Show all your working.

A There are many ways to work this out. One method is to work out the cost of 25 ml for each size of tube.

If **50 ml** costs £1.24, 25 ml = £1.24 ÷ 2

= **62p**

If **75 ml** costs £1.96, 25 ml = £1.96 ÷ 3

= **65.3̇p**

If **100 ml** costs £2.42, 25 ml = £2.42 ÷ 4

= **60.5p**

The 100 ml tube of toothpaste is the best value for money, since it is the cheapest per 25ml of toothpaste.

Exam tips You can use the unitary method to work out which is the best buy. Work out either the cost per unit weight (least cost is best buy) or the amount per £1 (greatest amount is the best buy).

Practice questions

1. Janis is ordering some wool from Turkey. It costs ₺8.29 (Turkish Lira) per ball and she orders six balls. ₺1 is worth £0.24. What is the total cost of the wool in pounds?

2. Copy and complete the table.

£1 = $1.25

£	$
10	
	15
25.50	
	48.40

3. You can buy a box of six crackers for £1.50 or a box of ten crackers for £2.25.

Which box is better value for money?

4 Terence is ordering some dog food online. He sees these two offers for the same brand of food. Which is better value?

5 The hand cream Bernice uses can be bought in three sizes.

Which is the best value for money?

13.4 Increasing and decreasing in a given ratio

To increase or decrease quantities in a given ratio, you can either:

• find the unit amount and then multiply

• work out the scale factor and multipy.

Example 12	**Q** A recipe for four people requires 1600 g of flour. How much flour is needed for six people?	
	A $1600 \div 4 = 400$ g of flour for one person 400 g $\times 6 = 2400$ g of flour for six people	**Alternative method:** 4 people 1600 g 2 people 800 g 6 people 2400 g

Example 13	**Q** This is a recipe for 8 biscuits.
	80 g butter
	100 g sugar
	2 eggs
	120 g flour
	Sophie has 190 g of flour. Is this enough to make 12 biscuits?

(continued)

A You can find the amount needed for one biscuit and then multiply.

120 g flour for 8 biscuits

$\frac{120}{8}$ = 15 g flour for 1 biscuit

12 × 15 g = 180 g flour for 12 biscuits

Yes, Sophie has enough flour to make 12 biscuits.

Alternatively, you can work out the **scale factor** or multiplier.

$\frac{12}{8}$ = 1.5, so 1.5 times as much flour is needed for 12 biscuits.

1.5 × 120 g = 180 g

Yes, Sophie has enough flour to make 12 biscuits.

When two quantities are in **inverse proportion**, one quantity increases at the same rate as the other quantity decreases. For example, the time it takes to build a wall increases as the number of builders decreases.

Example 14

Q It took four builders six days to build a wall. How long would it take six builders to build the same wall?

A 4 builders take 6 days, so 1 builder would take 6 × 4 = 24 days

Working at the same rate, it would take 6 builders $\frac{24}{6}$ = 4 days

Exam tips Always check that your answer to a proportion question sounds reasonable at the end. In the example above, you should expect more builders to take less time to build the wall.

Practice questions

1. Here is a recipe to make six potato cakes. Work out how much of each ingredient is needed to make nine potato cakes.

Potato cakes (Makes 6)

2 cups mashed potato
1 onion
1 egg
1 cup plain flour

2. A 10 kg sack of dry food will last Jay's dog for four weeks. How long will a 15 kg sack last?

3. Thierry is making bath fizzers. Each bath fizzer uses 100 g baking soda and 50 g citric acid. Thierry has 740 g baking soda and plenty of citric acid. What is the maximum number of bath fizzers that he can make?

4 This recipe makes 30 biscuits.

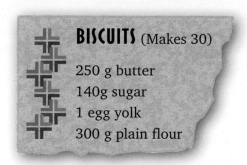

BISCUITS (Makes 30)

250 g butter

140g sugar

1 egg yolk

300 g plain flour

Piotr has 420 g plain flour but plenty of all the other ingredients. Does he have enough flour to make 45 biscuits?

5 Five dogs eat a 12 kg sack of food in 14 days. How long would the same sack of food last eight dogs?

REVISION CHECKLIST

- A ratio is used to compare two or more related quantities. Ratios can be used to solve problems involving best buys, exchange rates or quantities in recipes.

- Two quantities are in direct proportion when both quantities increase at the same rate.

- Two quantities are in inverse proportion when one quantity increases at the same rate as the other quantity decreases.

Exam-style questions

1 Which of these will 15 : 35 simplify to?

| 3 : 8 | 5 : 7 | 7 : 3 | 3 : 7 | 5 : 12 |

2 Simplify 400 g : 2 kg.

3 Write the ratio 40 : 75 in the form 1 : p.

4 Which of these is the ratio 64 : 100, written in the form m : 1?

| 1 : 6.4 | 6.4 : 1 | 0.64 : 1 | 1.5625 : 1 | 1 : 1.5625 |

5 Sharne and Stevie share £150 in the ratio 7 : 3.

How much will each one receive?

6 Julien works in an ice-cream parlour. Last week he used 15 tubs of chocolate ice cream, 9 tubs of strawberry ice cream and 6 tubs of vanilla ice cream. This week he orders 70 tubs in the same ratio as the tubs he used last week. How many tubs of vanilla ice cream does he order?

7 Rob and Paul need to drive 650 miles. They share the driving in the ratio 5 : 8 with Rob driving the lesser amount. How many miles does Paul drive?

8 Nessa and her grandma share the cost of an e-reader in the ratio 1 : 4 with her grandma paying the greater share. The e-reader costs £79. How much more does Nessa's grandma pay than Nessa?

9 Enid and Jill share some money in the ratio 5 : 8.
   a What fraction of the money will Enid receive?
   b What percentage of the money will Jill receive?

10 A quadrilateral has angles in the ratio 1 : 2 : 3 : 4. Calculate the size of the largest angle.

11 Lily and Lydia each buy the same lipstick. Lily pays £14 and Lydia pays $19. Use the conversion £1 = $1.25 to work out which one paid less.

12 Ben and Ken both order the new WorldCraft book. Ben orders it from the UK and pays £19.50. Ken orders it from the USA and pays $15.60. The conversion rate is £1 = $1.30.

   Who paid less for the book?

13 Maria went on holiday to Scotland from Italy. She exchanged €300 to pounds at an exchange rate of £1 = €1.20. How many pounds did she receive?

14 Henry wants to buy some jewels for the game is playing online. He can choose from these buying options:

   1 jewel = 50p

   5 jewels = £1.70

   10 jewels = £3.50

   Which of these offers the best value for money?

15 This recipe is for 30 biscuits.

   Piotr only has 200 g sugar, but an unlimited supply of the other ingredients. Does he have enough sugar to make 40 biscuits?

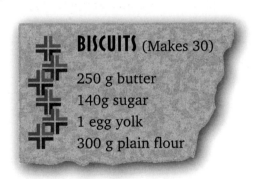

BISCUITS (Makes 30)

250 g butter

140g sugar

1 egg yolk

300 g plain flour

16 It takes 20 men six days to make 500 doors. How many doors could four men make in six days?

Now go back to the list of objectives at the start of this chapter.
How confident do you now feel about each of them?

14 Measurement

Objectives

Before you start this chapter, mark how confident you feel about each of the statements below:

	▶	▶▶	▶▶▶
I can use standard units of mass, length, time, money and other measures.			
I can read a timetable.			
I can use compound measures such as speed, density and pressure.			
I can use ratios to find the values of missing lengths on scales and maps.			
I can use compound measures such as rates of pay and unit pricing to solve problems in real-life contexts.			

Check-in questions

- Complete these questions to assess how much you remember about each topic. Then mark your work using the answers at the back of the book.
- If you score well on all sections, you can go straight to the Revision checklist and Exam-style questions at the end of the chapter. If you don't score well, go to the chapter section indicated and work through the examples and practice questions there.

1. Change 3200 g into kilograms. Go to 14.1

2. Write 3:30 a.m. as a 24-hour clock time. Go to 14.1

3. The scale on a map is 1 cm to 5 km. What distance is represented by a line of length 6.5 cm? Go to 14.2

4. Find the time taken for a car to travel 240 miles at an average speed of 70 mph. Go to 14.3

14.1 Units of measurement

The **metric system** of units is based on tens.

Metric units

Here are some metric units.

Length	Mass	Capacity
10 mm = 1 cm	1000 mg = 1 g	1000 ml = 1 l
100 cm = 1 m	1000 g = 1 kg	100 cl = 1 l
1000 m = 1 km	1000 kg = 1 tonne	1000 cm^3 = 1 l

- Divide to change from smaller units to larger units, e.g. grams → kilograms.
- Multiply to change from larger units to smaller units, e.g. metres → centimetres.

Exam tips When converting from one unit to another, first decide whether your answer should be larger or smaller, then multiply or divide as appropriate.

Example 1

Q Convert the following.

  a 500 cm to metres   **b** 5 litres to centilitres

  c 3500 g to kilograms   **d** 25 cm to millimetres

A **a** 500 cm = 5 m (÷ 100)   **b** 5 litres = 500 cl (× 100)

    c 3500 g = 3.5 kg (÷ 1000)   **d** 25 cm = 250 mm (× 10)

When measuring it is important to choose sensible units.

Thickness of a book	Capacity of a bottle of water	Width of a bookcase	The amount of water in a swimming pool	Height of a house
millimetres	centilitres	centimetres	litres	metres
Mass of a packet of crisps	Length of a river	Mass of a bag of potatoes	Amount of medicine in a spoonful	Mass of a lorry
grams	kilometres	kilograms	millilitres	tonnes

Comparing metric and imperial units

Here are some comparisons between metric and imperial units. Note that these comparisons are only approximate.

Length	Mass	Capacity
2.5 cm ≈ 1 inch	28 g ≈ 1 ounce	1 litre ≈ $1\frac{3}{4}$ pints
8 km ≈ 5 miles	1 kg ≈ 2.2 pounds	4.5 litres ≈ 1 gallon

Measuring time and reading timetables

You can read or give times using either the 12-hour or 24-hour clock format.

For the 12-hour clock, times before midday are a.m. and times after midday are p.m.

The 24-hour clock numbers the hours from 0 to 24 so times are written using four digits.

Example 2

Q Convert these times to 24-hour clock times.

    a 2:42 p.m.   **b** 5:30 a.m.

A **a** 14:42   **b** 05:30

Example 3

Q Convert these to 12-hour clock times.

    a 15:27   **b** 07:04

A **a** 3:27 p.m.   **b** 7:04 a.m.

Exam tips 12 midnight is 12 a.m.
12 noon or midday is 12 p.m.

24-hour clock times often appear on bus and train timetables.

You will also need to be able to convert between different units of time.

- There are 60 seconds in one minute.
- There are 60 minutes in one hour.
- There are 24 hours in one day.
- There are seven days in one week.
- There are 52 weeks in one year.
- There are 365 days in a year, or 366 in a leap year.

Example 4

Q The timetable gives the times of some of the trains from London to Manchester.

London Euston	0702	0740	Every 60 minutes until	1100	1400
Watford Junction	0732	0812		1130	1430
Stoke-on-Trent	0850	0930		--	1545
Manchester Piccadilly	0940	1015		1315	1640

The 0850 train from Stoke-on-Trent.

The 0740 train from London Euston arrives at 1015.

The 1100 train from London Euston does not stop at Stoke-on-Trent.

a Jade travels from Watford Junction to Manchester Piccadilly to attend a university open day. She catches the 0732 train from Watford. How long should the train take to travel to Manchester Piccadilly?

b Colin arrives at London Euston at 1242. How long should he have to wait for the next train to Manchester?

A a Departs Watford at 0732, arrives in Manchester at 0940

Her journey time is 2 hours 8 minutes.

b 1242 → 1300 = 18 minutes

1300 → 1400 = 1 hour

He must wait 1 hour 18 minutes.

Exam tips Remember, for example, that 3.5 hours = 3 hours 30 minutes. To convert decimal parts of an hour to minutes, multiply by 60. For example, 0.3 hours is 0.3 × 60 = 18 minutes.

Practice questions

1 Suggest a sensible metric unit to measure the:

a mass of a large car

b height of a door

c depth of a washing machine

d thickness of a car tyre

e volume of drink in a bottle.

2 Copy and complete this table.

1 cm	=	10 mm	=	0.01 m
86 cm	=	___ mm	=	___ m
1 litre	=	1000 ___	=	100 ___
56 kg	=	___ tonnes	=	56 000 ___
206 m	=	20 600 ___	=	___ km

3 There are 12 inches in one foot. Jeff is 6 ft 2 inches tall. How tall is this in centimetres?

4 Copy and complete the table of 12-hour and 24-hour clock conversions.

12-hour clock time	24-hour clock time
1:58 p.m.	
	17:43
2:06 a.m.	
	11:30
9:15 a.m.	
	16:49

5) Hilda is going to a meeting starting in Manchester at 12 noon. It will take her 25 minutes to walk from the station at Manchester to the meeting venue. What is the time of the latest train she can catch from London Euston to arrive at her meeting on time?

London Euston	0900	0920	0940	1000
Milton Keynes		0950		
Stoke-on-Trent	1025	1048		1125
Crewe			1111	
Macclesfield	1041			1141
Wilmslow			1127	
Stockport	1055	1117	1136	1155
Manchester Piccadilly	1107	1128	1148	1207

14.2 Maps and diagrams

Scales are often used on maps and diagrams. They are usually written as **ratios**.

Example 5	Q Some students are taking part in a fun run for charity. They are running between marker A and marker B. The scale on a road map is 1 : 25 000. Marker A and marker B are 60 cm apart on the map. Work out the actual distance between them in kilometres.
	A A scale of 1 : 25 000 means that 1 cm on the map represents 25 000 cm in real life. Therefore, 60 cm represents 60 × 25 000 = 1 500 000 cm.
	1 500 000 ÷ 100 = 15 000 m ●—— Divide by 100 to change cm to m.
	15 000 ÷ 1000 = 15 km ●—— Divide by 1000 to change m to km.
	The actual distance between marker A and marker B is 15 km.

Example 6	Q A house plan has a scale of 1 : 30. If the width of the house on the plan is 64 cm, what is the width of the actual house?
	A A scale of 1 : 30 means that 1 cm on the plan represents 30 cm on the house. Therefore, 64 cm represents 64 × 30 = 1920 cm.
	1920 ÷ 100 = 19.2 m
	The width of the actual house is 19.2 m.

See Chapter 13 for more work on ratios.

Exam tips | Always look very carefully at the scales on maps and diagrams.

Practice questions

1 The scale on a map is 1 cm to 10 km. What distance is represented by a line of length 6 cm on the map?

2 This map shows an area of England and Wales.

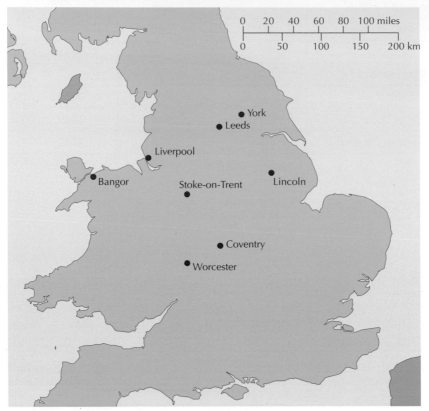

Which pair of towns is closest together? You must show all your working.

a Liverpool and York

b Worcester and Leeds

c Coventry and Bangor

d Lincoln and Stoke-on-Trent

3 Find a map showing motorways in England.

Use a piece of string to measure these map distances and then work out the actual distances. State the scale on the map and the measured distance in each case.

a Nottingham to Leicester using the M1

b Leicester to Coventry using the M69

c Coventry to Stoke-on-Trent using the M6

4 A map is drawn to a scale of 1 cm to 5 miles. Copy and complete the table below.

Actual distance	Distance on map
___ miles	2 cm
70 miles	___ cm
180 miles	___ cm

14.3 Compound measures

Compound measures involve two other measures of different types. You need to understand and be able to use compound measures related to speed, pressure and density.

You should work in steps when you convert between these compound measures.

Speed

$$\text{Speed}\ (s) = \frac{\text{distance travelled}\ (d)}{\text{time taken}\ (t)}$$

This can be rearranged to give:

$$\text{Time taken} = \frac{\text{distance travelled}}{\text{speed}}$$

$$\text{Distance travelled} = \text{speed} \times \text{time taken}$$

Units of speed include:

- metres per second (m/s)
- kilometres per hour (km/h)
- miles per hour (mph).

Example 7	**Q** Change 80 km/h into metres/second.
	A 80 km in 1 hour $=$ 80 000 m in 1 hour
	$=$ 80 000 m in 60 minutes
	$=$ 1333.$\dot{3}$ m in 1 minute
	$=$ 22.$\dot{2}$ m in 1 second

Exam tips Remember to show each step of your working and state the units.

Example 8	**Q** A car travels 80 km in 1 hour 20 minutes. Find the speed in kilometres/hour.
	A $s = \dfrac{d}{t}$
	$= \dfrac{80}{1.\dot{3}}$ ← Change the time into hours. 20 minutes is $\frac{20}{60}$ of 1 hour.
	$= 60$ km/h

Example 9	**Q** Josie is driving her car at a speed of 70 mph.
	Audrey is driving her car at a speed of 120 km/h.
	Given that 5 miles ≈ 8 km, who is travelling faster? Show working to support your answer.
	A Either convert Josie's speed to km/h or Audrey's speed to mph.
	Josie's speed in km/h is $70 \times \dfrac{8}{5} = 112$ km/h
	Audrey is travelling at 120 km/h, so she is travelling faster.

Example 10

Q Miss Fitzgerald drives 40 miles to work. On Wednesday her journey to work took 50 minutes. On Thursday the average speed of her journey to work was 54 km/h.

Did Miss Fitzgerald drive to work more quickly on Wednesday or Thursday?

A Speed on Wednesday is $40 \div \frac{50}{60} = 48$ mph

1 km $\approx \frac{5}{8}$ mile, so speed on Thursday is $\frac{5}{8} \times 54 = 33.75$ mph.

Miss Fitzgerald drove more quickly to work on Wednesday.

Density

Density is the mass per unit volume of any object.

$$\text{Density } (D) = \frac{\text{mass } (m)}{\text{volume } (v)} \qquad \text{Volume} = \frac{\text{mass}}{\text{density}} \qquad \text{Mass} = \text{volume} \times \text{density}$$

Example 11

Q Find the density of an object with a mass of 600 g and a volume of 50 cm^3.

A Density $= \dfrac{\text{mass}}{\text{volume}}$

$= \dfrac{600}{50}$

$= 12$ g/cm^3

Example 12

Q Two solids each have a volume of 2.5 m^3.

The density of solid A is 320 kg/m^3.

The density of solid B is 288 kg/m^3.

Calculate the difference in the masses of the solids.

A
$$\text{Mass} = \text{volume} \times \text{density}$$
Mass of solid A $= 2.5 \times 320 = 800$ kg

Mass of solid B $= 2.5 \times 288 = 720$ kg

Difference in mass $= 800 - 720$

$= 80$ kg

Pressure

Pressure is defined as force per unit area.

$$\text{Pressure } (P) = \frac{\text{force on surface } (F)}{\text{surface area } (A)}$$

$$\text{Force on surface } (F) = \text{pressure } (P) \times \text{surface area } (A)$$

$$\text{Surface area } (A) = \frac{\text{force on surface } (F)}{\text{Pressure } (P)}$$

| Example 13 | **Q** A force of 17 N acts over an area of 5 m². What is the pressure? |

A $P = \dfrac{F}{A}$

$= \dfrac{17}{5}$

$= 3.4 \text{ N/m}^2$

Exam tips Use a formula triangle to help you remember the formulae for speed, density and pressure.

Speed Density Pressure

Rates of pay and rates of usage

A rate is also a compound unit. It tells you how many units of one quantity there are compared with one unit of another quantity. For example, rate of pay $= \dfrac{\text{amount of money}}{\text{time}}$.

| Example 14 | **Q** Fiona is paid £502 a week. Each week she works 40 hours. What is her hourly rate of pay? |

A Rate of pay $= \dfrac{£502}{40 \text{ hours}}$

$= £12.55 \text{ per hour}$

| Example 15 | **Q** An electric heater uses 20 units of electricity over 8 hours. What is the hourly rate of electricity consumption in units per hour? |

A Hourly rate $= \dfrac{20 \text{ units}}{8 \text{ hours}}$

$= 2.5 \text{ units per hour}$

| Example 16 | **Q** Water empties from a tank at a rate of 1.75 litres per second.

It takes 10 minutes to empty the tank.

How much water was in the tank? |

A 10 minutes $= 10 \times 60$

$= 600 \text{ seconds}$

Volume of water in the tank $= 1.75 \times 600$

$= 1050 \text{ litres}$

Practice questions

1 Copy and complete this table.

	Mass (g)	Density (g/cm³)	Volume (cm³)
a	3.6		4.5
b	54	1.2	
c		3.15	60

2 Copy and complete this table.

	Force	Pressure	Area
a		20 N/m²	20 m²
b	106 N		212 cm²
c	50 N	1.25 N/cm²	

3 The density of iron is 7900 kg/m³. Work out the mass of 4.2 m³ of iron.

4 Copy and complete this table.

Speed	Distance	Time
100 m/s	250 m	___ s
60 mph	___ miles	3.5 hours
___ mph	450 miles	$\frac{1}{2}$ hour

REVISION CHECKLIST

Length	Mass	Capacity
10 mm = 1 cm	1000 mg = 1 g	1000 ml = 1 l
100 cm = 1 m	1000 g = 1 kg	100 cl = 1 l
1000 m = 1 km	1000 kg = 1 tonne	1000 cm³ = 1 l

- 12-hour clock times are followed by a.m. before midday and by p.m. after midday.

- 24-hour clock times have four digits. The hours are numbered from 00 to 24.

- Speed $= \dfrac{\text{distance travelled}}{\text{time taken}}$

- Density $= \dfrac{\text{mass}}{\text{volume}}$

- Pressure $= \dfrac{\text{force}}{\text{surface area}}$

- Rate of pay $= \dfrac{\text{amount of money}}{\text{time}}$

Exam-style questions

1   Dan wants to buy a case for his tablet. He sees one for sale online advertised with the dimensions $16 \times 25 \times 1.5$ mm. Dan says this must be wrong. Explain why.

2   Write 1 mm in metres.

3   What is 560 cm^3 in litres?

0.56 litres	56 litres	560 000 litres	5.6 litres

4   Gladys is travelling from London to Stockport by train. Jim wants to get on the same train at Stoke-on-Trent. If they want to arrive into Stockport no later than 1130, what is the time of the latest train Gladys can catch from London?

London Euston	0900	0920	0940	1000
Milton Keynes		0950		
Stoke-on-Trent	1025	1048		1125
Crewe			1111	
Macclesfield	1041			1141
Wilmslow			1127	
Stockport	1055	1117	1136	1155
Manchester Piccadilly	1107	1128	1148	1207

5   What is 1:40 a.m. written as a 24-hour clock time?

1:40	13:40	11:40	01:40	10:40	1.40pm

6   Write these lengths in order of size, starting with the smallest.

5.6 km   78 mm   650 m   4589 cm   3.5 m

7   Beverly has a full 2-litre carton of juice. She drinks 170 ml from the carton. How much is left?

8   A map is drawn to a scale of 1 cm to 10 miles. Copy and complete the table below.

Actual distance	Distance on map
___ miles	8 cm
60 miles	___ cm
1000 miles	___ m

9   A map uses a scale of 1 cm to 5 km.

   a Two places are 15 cm apart on the map. What is the actual distance between them?

   b Two places are 90 km apart. What is the distance between them on the map?

10   A scale drawing uses a scale of 1 cm to 3 m. The actual length AB is 42 m. What is the length of AB in the diagram?

11 The scale of a map is 1 : 200 000.

    a Two places are 15 cm apart on the map. What is the actual distance between them?

    b Two places are 71 km apart. How is this represented on the map?

12 Sara is looking at a map of the area she will visit on holiday. The map scale is 1 cm to 5 km.

    She plans to walk a path that is 24 km long. What is the length of the path on the map?

13 Copy and complete the table.

Speed	Distance	Time
100 m/s	350 m	___ s
65 mph	___ miles	3.5 hours
___ mph	150 miles	20 minutes

14 Deji walks 2 km in 15 minutes.

    a Work out her speed in kilometres per hour.

    b How long would it take her to walk 25 km at this speed?

15 A car is travelling at an average speed of 40 mph. ●——[See Chapter 11.]

    a Create a table of values and then draw a distance–time graph.

    b What distance will the car travel in $2\frac{1}{2}$ hours?

16 Clara runs at an average speed of 12 km/h. How many minutes does it take her to run 6 km?

17 Calculate the distance travelled in 2 hours at a constant speed of 80 km/h.

18 A cube of side 20 cm exerts a force of 100 N on a table. What is the pressure on the table?

19 It takes a plane 50 minutes to fly 450 miles from London to Edinburgh. What is the average speed of the aeroplane?

20 The density of tin is 7260 kg/m^3. Find the mass of 16 m^3 of tin. Give your answer in kilograms.

Now go back to the list of objectives at the start of this chapter.

How confident do you now feel about each of them?

15 Similarity and congruence

Check-in questions

- Complete these questions to assess how much you remember about each topic. Then mark your work using the answers at the back of the book.

- If you score well on all sections, you can go straight to the Revision checklist and Exam-style questions at the end of the chapter. If you don't score well, go to the chapter section indicated and work through the examples and practice questions there.

1. Calculate the lengths marked *n* in these pairs of similar shapes. Give your answers correct to one decimal place.   Go to 15.1

a
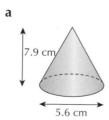
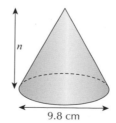

b
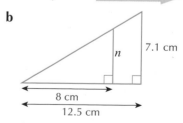

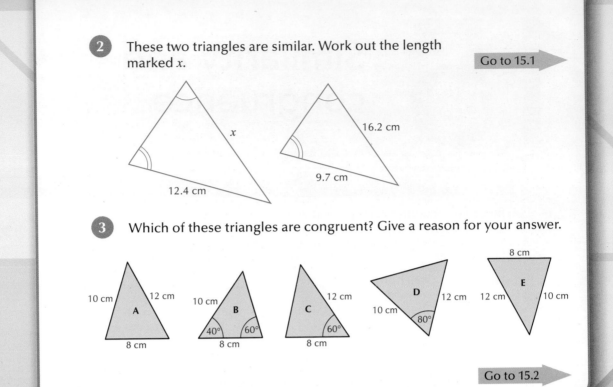

2 These two triangles are similar. Work out the length marked x.

Go to 15.1

16.2 cm

9.7 cm

x

12.4 cm

3 Which of these triangles are congruent? Give a reason for your answer.

8 cm

10 cm 12 cm

A

8 cm

10 cm

B

40° 60°

8 cm

12 cm

C

60°

8 cm

D

10 cm 12 cm

80°

E

12 cm 10 cm

Go to 15.2

15.1 Similar shapes

Objects that are exactly the same shape but different sizes are called **similar** shapes. One shape is an enlargement of the other. The corresponding angles in the shapes are equal; the corresponding lengths are in the same ratio.

These shapes are similar. Each side of the first triangle is twice the length of the corresponding side of the second triangle. In other words, the second triangle is an enlargement of the first triangle with a scale factor of 2.

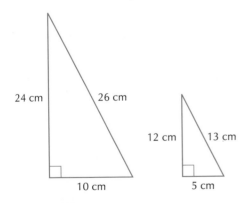

24 cm 26 cm

10 cm

12 cm 13 cm

5 cm

Finding unknown lengths in similar figures

You need to be able to calculate the lengths of unknown sides in similar shapes. You can do this by using the ratio of corresponding sides.

Example 1

Q In each diagram, find the value of the length marked a.

a

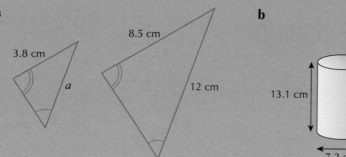

b

A a Corresponding lengths are in the same ratio, so:

$$\frac{a}{12} = \frac{3.8}{8.5}$$

$$a = \frac{3.8}{8.5} \times 12$$ ●——— Multiply both sides by 12.

$$= 5.36 \text{ cm (3 s.f.)}$$

b $$\frac{a}{7.2} = \frac{19.5}{13.1}$$

$$a = \frac{19.5}{13.1} \times 7.2$$ ●——— Multiply both sides by 7.2.

$$a = 10.7 \text{ cm (3 s.f.)}$$

Example 2

Q Calculate the length marked y.

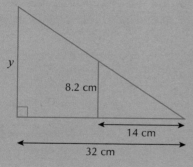

A Redraw the diagram as two separate triangles.

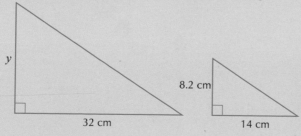

$$\frac{y}{8.2} = \frac{32}{14}$$

$$y = \frac{32}{14} \times 8.2$$ ●——— Multiply both sides by 8.2.

$$y = 18.7 \text{ cm (3 s.f.)}$$

Practice questions

1 The diagram shows a picture inside a picture frame. Are the two rectangles similar? Explain your answer.

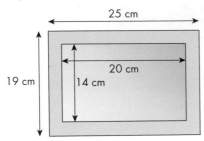

2 These two triangles are similar.

   a Explain how you can calculate the unknown length in the larger triangle.

   b Explain how you can find the values of the unknown angles in the smaller triangle.

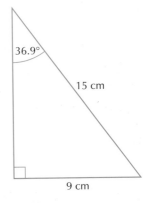

3 The diagram shows two similar triangles. Work out the values of x, y and z.

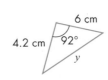

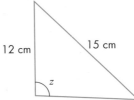

4 Eowyn has spilled his drink over his book. He knows the two triangles are similar. He says he is confident he can write down the value of x, but not y. Explain why.

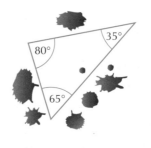

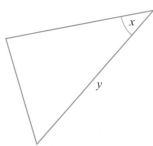

5 These two triangles are similar. Calculate the values of the unknown lengths and angles in both triangles.

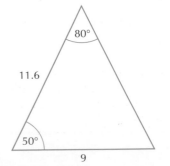

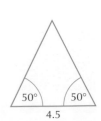

15.2 Congruent shapes

Shapes are **congruent** if they are exactly the same size and shape, i.e. they are identical. Two shapes are congruent even if they are in a different orientation or mirror images of each other.

In order for a pair of triangles to be congruent, they must meet one of the sets of conditions listed in the table.

Condition	Description	Example
SSS (Side–Side–Side)	The three sides of one triangle are the same as the three sides of the other triangle.	
SAS (Side–Angle–Side)	Two sides and the angle between them in one triangle are the same as two sides and the angle between them in the other triangle.	
RHS (Right Angle–Hypotenuse–Side)	Each triangle contains a right angle. Both hypotenuses and another pair of sides are equal.	
AAS (Angle–Angle–Side)	Two angles and a side in one triangle are equal to two angles and the corresponding side in the other.	

Example 3

Q CDE is an equilateral triangle. F lies on DE. CF is perpendicular to DE.

Prove that triangle CFD is congruent to triangle CFE.

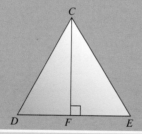

A There are several ways of proving this. For example:

Since triangle CDE is equilateral, CD = CE. CD is the hypotenuse of triangle CFD and CE is the hypotenuse of triangle CFE.

CF is common to triangle CFD and triangle CFE.

Angle CFD = angle CFE = 90°.

So triangle CFD is congruent to triangle CFE by RHS.

Exam tips When proving that triangles are congruent, you must give reasons for each stage of your proof. Remember to state which rule your proof satisfies.

Practice questions

1 List all the sets of congruent triangles.

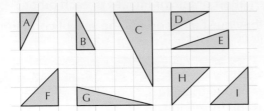

2 Identify the pairs of congruent triangles. State the condition for congruency.

Diagrams not drawn accurately.

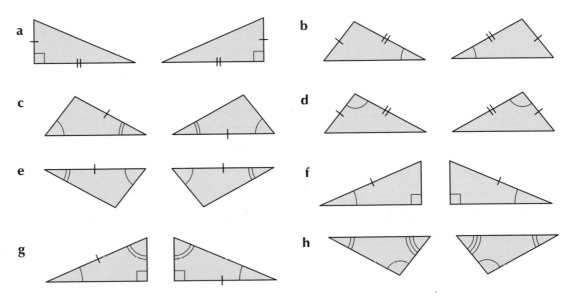

a

b

c

d

e

f

g

h

3 How can you tell that triangle ABC and triangle DEC are congruent?

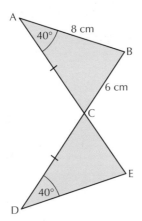

4 **a** How do you know that triangle ABC and triangle DEC are congruent?

b Work out the lengths CD and CE.

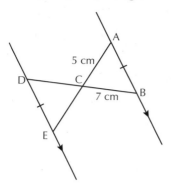

5   The diagram shows a rhombus.

a On a copy of the diagram, mark any equal sides, parallel lines and equal angles.

b Write down any congruent triangles.

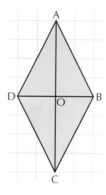

REVISION CHECKLIST

● Objects that are exactly the same shape but not the same size are called similar shapes.

● In similar shapes, corresponding angles are equal.

● In similar shapes, corresponding lengths are in the same ratio.

● Shapes are congruent if they are exactly the same size and shape (i.e. they are identical).

● The conditions of congruence for triangles are SSS, SAS, RHS and AAS or ASA.

Exam-style questions

1   M and N are similar shapes. What is the scale factor from M to N?

| 4 | 0.25 | 3 | 2.4 | 1.5 |

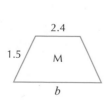

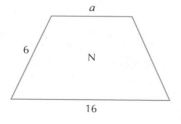

2 Which shape is congruent to shape X?

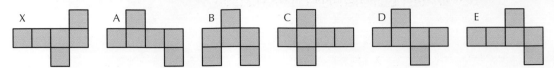

3 M and N are similar shapes. What is the value of a?

0.6	3	4	9.6	24

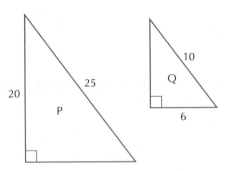

4 P and Q are similar triangles.

What is the scale factor from P to Q?

2.5	2	15	0.4	4

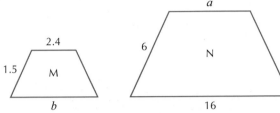

5 P and Q are similar triangles.

Write down the value of the unknown length for each triangle.

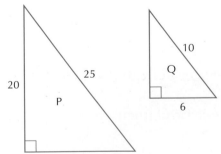

6 XYZ and ABC are similar triangles where AB = 2XY.

Say whether each statement is true or false.

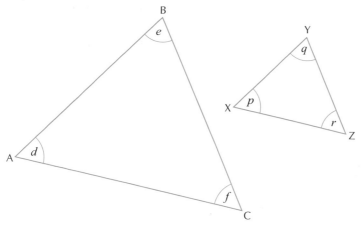

a $d = 2$ **b** XZ = 2AC   **c** $e = \dfrac{1}{2}$ **d** BC = qr

7 Rectangles P and Q are similar.

   a Calculate the value of x.

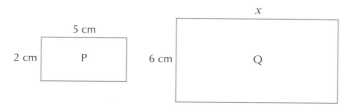

   b The area of rectangle P is 10 cm². Aditya says that the area of rectangle Q must be 30 cm². Is he correct? You must explain your answer.

8 ABC and XYZ are similar shapes.

   a Work out the length of XZ.

   b Work out the length of BC.

   c Which one of these statements is true?

      i Angle B is twice the size of angle Y.

      ii Angle B is the same size as angle Y.

      iii We do not have enough information to compare angles B and Y.

9 These three rectangles are similar.

   a Work out the length of a.

   b Work out the length of b.

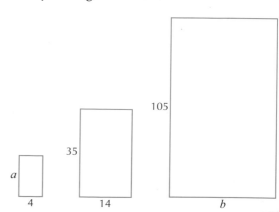

Triangles

All triangles have three sides. There are different types, defined by side and angle properties.

Right-angled

- Has a 90° angle

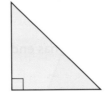

Equilateral

- Three equal sides
- Three equal angles
- Three lines of symmetry

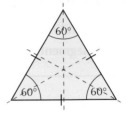

Isosceles

- Two equal sides
- Two equal angles
- One line of symmetry

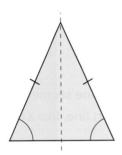

Scalene

- No equal sides or equal angles
- No lines of symmetry

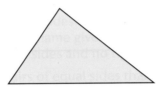

Quadrilaterals

Quadrilaterals have four sides. There are different types, defined by the properties of their angles and their sides.

Square

- Four lines of symmetry
- Rotational symmetry of order 4
- All sides equal
- Two pairs of parallel sides
- All angles are 90°
- The diagonals are equal and bisect each other at right angles.

Rectangle

- Two lines of symmetry
- Rotational symmetry of order 2
- Opposite sides equal
- Two pairs of parallel sides
- All angles 90°
- The diagonals bisect each other.

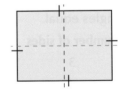

Rhombus

- Two lines of symmetry
- Rotational symmetry of order 2
- All sides equal
- Opposite sides parallel
- Opposite angles equal
- The diagonals bisect each other at right angles and also bisect the corner angles.

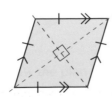

Parallelogram

- No lines of symmetry
- Rotational symmetry of order 2
- Opposite sides equal and parallel
- Opposite angles equal

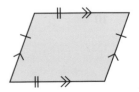

Kite

- One line of symmetry
- No rotational symmetry
- Two pairs of adjacent sides are equal.
- The diagonals do not bisect each other but cross at right angles.

Trapezium

- One pair of parallel sides
- No lines of symmetry
- No rotational symmetry
- **Special case:** An isosceles trapezium has one pair of equal sides and one line of symmetry.

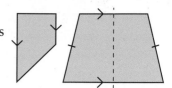

Practice questions

1. Draw a sketch of an isosceles triangle. Mark on any important features.

2. The triangle in the diagram is not drawn accurately.

   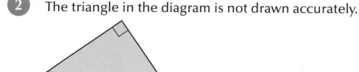

   Which of these statements could be true?

   a The triangle is isosceles.

   b The triangle is equilateral.

   c The triangle is scalene.

3. Erin draws this regular hexagon.

   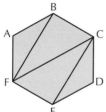

   a Identify as many sets of parallel lines as you can.

   b Erin says triangle CDE is an isosceles triangle. Is she correct? You must explain your answer.

   c What is the mathematical name of the quadrilateral formed by ABCF?

4. Which of these statements are true and which are false?

   a All rectangles are squares.

   b All squares are rectangles.

   c A parallelogram is a special type of rectangle.

   d A rectangle is a special type of parallelogram.

   e A rhombus is a special type of parallelogram.

   f A parallelogram is a special type of rhombus.

5. James draws a quadrilateral. The diagonals of the quadrilateral bisect each other. List the quadrilaterals that he could have drawn.

6. Mary draws a quadrilateral. The diagonals of the quadrilateral are perpendicular to each other. List the quadrilaterals that she could have drawn.

16.2 Symmetry

You will need to know about two different types of symmetry in 2D shapes.

Reflection symmetry

When a 2D shape has **reflection symmetry** or line symmetry, a mirror line can be drawn where one half of the shape is the reflection of the other half. The mirror line is called the **line of symmetry**. These are examples of line symmetry in shapes.

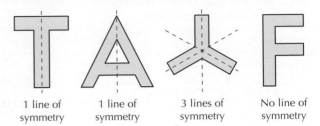

| 1 line of symmetry | 1 line of symmetry | 3 lines of symmetry | No line of symmetry |

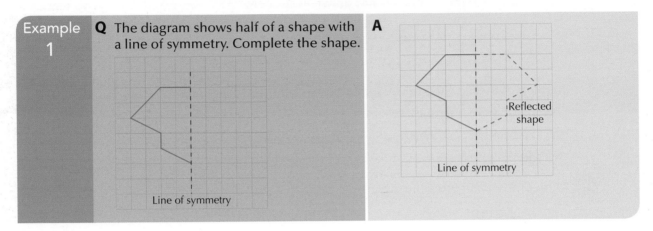

| Example 1 | **Q** The diagram shows half of a shape with a line of symmetry. Complete the shape. | **A** |

Line of symmetry

Reflected shape

Line of symmetry

Rotational symmetry

A 2D shape has rotational symmetry when it looks exactly the same after a rotation of less than one full turn (360°).

The **order of rotational symmetry** is the number of times the shape will look the same as it is rotated through one full turn. Here are some examples.

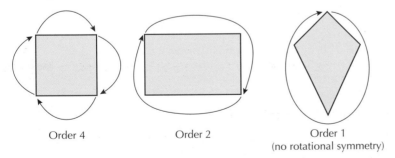

Order 4   Order 2   Order 1
(no rotational symmetry)

The kite must rotate through one full turn before it looks the same. We say it has rotational symmetry of order 1 or no rotational symmetry.

A regular pentagon has five equal sides, rotational symmetry of order 5 and 5 lines of symmetry.

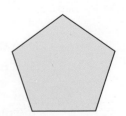

Exam tips When marking lines of symmetry or completing a shape to make it symmetrical, make sure that you draw accurately.

Practice questions

1 Draw a sketch of a regular octagon. Mark all the lines of symmetry.

2 Jane says that a rhombus has the same number of lines of symmetry as a rectangle. Is she correct? Draw diagrams to support your answer.

3 Copy this design onto squared paper. Complete the design by shading exactly three squares so that the design has exactly one line of symmetry.

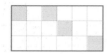

4 Phil says a regular pentagon has 5 lines of symmetry and rotational symmetry of order 5, so a regular hexagon must have 6 lines of symmetry and rotational symmetry of order 6. Is James correct? Draw diagrams to support your answer.

5 In each shape, the dashed line is a line of symmetry. Copy each shape onto squared paper and then complete it.

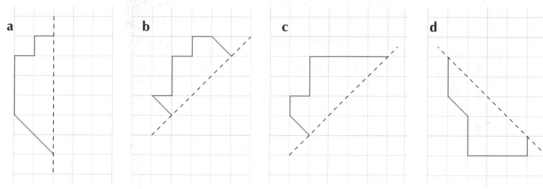

6 For each shape, write down:
  i the number of lines of symmetry
  ii the order of rotational symmetry.

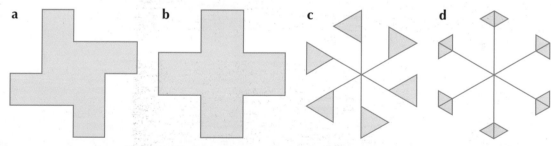

> **Exam tips** A regular polygon with n sides will have n lines of symmetry and order of rotational symmetry n.

16.3 Properties of 3D shapes

3D (three-dimensional) shapes are sometimes called solids. In exam questions involving 3D shapes, you need to be able to work out length, surface area and volume. You also need to identify properties of a range of 3D shapes.

See also Chapter 21.

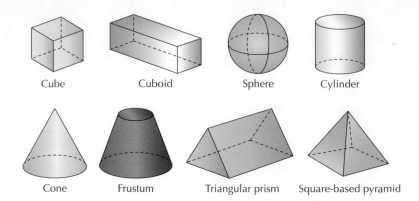

Cube   Cuboid   Sphere   Cylinder

Cone   Frustum   Triangular prism   Square-based pyramid

A **prism** is a 3D shape with a uniform **cross section** and flat faces. This means that if it were cut into slices along the cross section, the face of each slice would be the same shape and size.

A **face** is a flat surface of a solid.

An **edge** is where two faces meet.

Vertex is another word for corner.

A cuboid has 6 faces, 8 vertices and 12 edges.

A **square-based pyramid** has 5 faces and 8 edges.

The **net** of a 3D shape is a pattern of 2D (flat) shapes that can be folded to make the 3D solid. For example, the diagram shows a net for a cuboid.

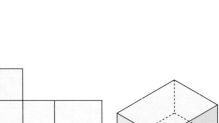

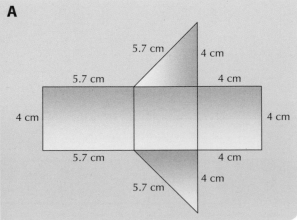

| Example 2 | **Q** Draw an accurate net of this 3D shape. 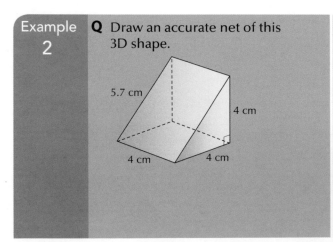 | **A** |

Exam tips   In an exam, you may be asked to draw an accurate net. This means you should measure all the lengths carefully and label them. Always check that the edges which will meet when the shape is folded have the same length.

Practice questions

1 Copy and complete the table to show the number of faces, edges and vertices for each shape.

Shape		Number of faces	Numbers of edges	Number of vertices
Cube				
Triangular prism				
Square-based pyramid				
Tetrahedron				

2 Which of these diagrams shows the net of a tetrahedron?

a   b   c   d   e

3 There are 11 different nets for a cube. Draw as many as you can on squared paper.

4 Draw a net for each prism.

a

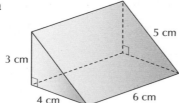

3 cm   5 cm   4 cm   6 cm

b

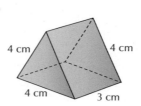

4 cm   4 cm   4 cm   3 cm

5 The diagram shows a net for a hexagonal prism.

How many vertices, faces and edges does the prism have?

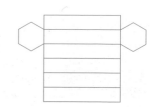

16.4 Plans and elevations

A **plan** is a 2D view of a 3D solid from above.

An **elevation** is a 2D view from the side or front of the shape.

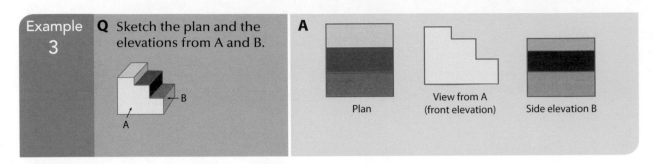

Example 3
Q Sketch the plan and the elevations from A and B.

A
Plan | View from A (front elevation) | Side elevation B

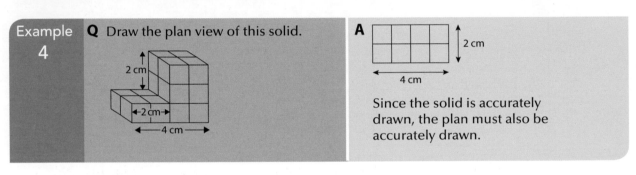

Example 4
Q Draw the plan view of this solid.

A
2 cm

4 cm

Since the solid is accurately drawn, the plan must also be accurately drawn.

Practice questions

1. The diagrams show models made from centimetre cubes. The number of cubes used to make each model is shown above the diagram.

   Draw the plan view and the elevation from the arrow for each diagram. Do not show individual cubes.

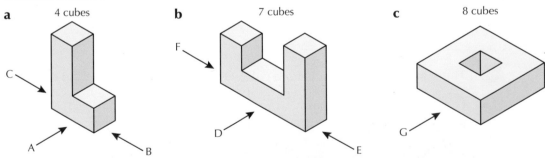

   a 4 cubes   **b** 7 cubes   **c** 8 cubes

2. Draw a sketch of the plan and elevations from X and Y.

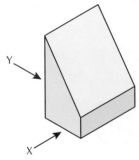

3 The diagram shows a model of a house. Draw the plan view and the elevation seen from each of the arrows.

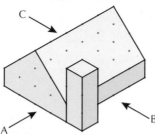

REVISION CHECKLIST

- A line segment has a finite length. For example, the line segment PQ has endpoints P and Q.

- Two lines are parallel in 2D if they do not intersect or touch each other. They are always equidistant (the same distance apart) from each other.

- Two lines are perpendicular if they are at right angles to each other. Perpendicular lines meet at 90°.

- When a 2D shape has reflection symmetry or line symmetry, a mirror line can be drawn so that one half of the shape is the reflection of the other half. The mirror line is called the line of symmetry.

- A 2D shape has rotational symmetry when it looks exactly the same after a rotation of less than one full turn (360°). The order of rotational symmetry is the number of times the shape will look the same as it is rotated through one full turn.

- A plan view is a 2D view from above the 3D shape.

- An elevation is a 2D view of a 3D shape from the side or front.

- The net of a 3D shape is a pattern of 2D (flat) shapes that can be folded to make the solid.

Exam-style questions

1 What is the mathematical name of this triangle?

2 The diagram shows a regular octagon.
  a Identify any sets of parallel lines.
  b What type of triangle is triangle ABC?
  c What type of quadrilateral is EFGH?
  d What type of quadrilateral is ADEH?

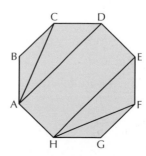

3 The diagram shows a rectangle.

    a Make an accurate drawing of the rectangle and mark on any lines of symmetry.

    b What is the order of rotational symmetry of your rectangle?

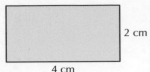

2 cm

4 cm

4 Write down **one** letter from those shown which has:

    a exactly two lines of symmetry

    b exactly one line of symmetry

    c no lines of symmetry

    d rotational symmetry of order 2.

5 Draw a quadrilateral which has exactly two lines of symmetry **and** diagonals which are perpendicular to each other.

6 Make a sketch of a rhombus and mark all its mathematical properties.

7 The diagram shows a parallelogram.

    a On a copy the diagram, mark any important mathematical features.

    b Describe any symmetrical properties.

8   **a** What is the mathematical name for this shape?

    b What is the order of rotational symmetry?

9 Make a sketch of a kite. Describe any mathematical features, including any equal sides, symmetrical properties and diagonals.

10 Rachel is making pendant earrings that she will sell to raise money for a charity.

    She decides that they will be in the shape of a heptagon and have exactly one line of symmetry.

    Draw a sketch of a design that she could use.

11 Sam, Jon, and Dan are each given two identical isosceles triangles.

    They each place their two triangles together to make a different quadrilateral.

    a Sketch the three quadrilaterals they make.

    b Write the mathematical name next to each quadrilateral.

    c On each quadrilateral, mark any equal sides, parallel sides and lines of symmetry.

    d What is the order of rotational symmetry of each quadrilateral?

12 Here are some properties of quadrilaterals.

    A Four lines of symmetry   **B** Two lines of symmetry

    C No lines of symmetry   **D** Rotational symmetry of order 4

    E Rotational symmetry of order 2   **F** No rotational symmetry

    G No parallel lines   **H** One pair of parallel sides

    I Two pairs of parallel sides

a Lilly draws a trapezium which has sides of different lengths. Which three properties does her trapezium have?

b Sketch a quadrilateral which has properties **B**, **E** and **I**. Show the lines of symmetry on your diagram.

13 The diagram shows a cuboid 5 cm by 3 cm by 2 cm.

Draw an accurate net for the cuboid.

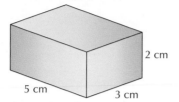

14 Here is the net for a prism.

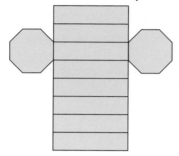

a What is the name of the polygon that forms the cross section of this prism?

b How many vertices, edges and faces does this prism have?

15 Copy and complete the following design so that it has **exactly** two lines of symmetry. Shade the smallest number of squares possible.

16 Copy and complete the following design by shading in seven more squares so that it has rotational symmetry of order 2.

17 The diagram shows an octahedron.

a Write down the number of
  i faces   **ii** edges   **iii** vertices.

b Sketch one possible net for this shape.

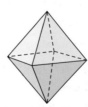

18 The diagram shows a 3D shape made from two identical tetrahedra glued together.

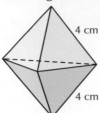

4 cm

4 cm

   a How many vertices, edges and faces does it have?

   b Sketch one possible net for the shape.

19 The diagrams show the plan view and end elevation of a solid.

   a Draw the side elevation.

   b Sketch the solid.

20 The diagrams show the plan and side elevation of a solid shape.

Side
elevation

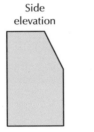

Plan

   a Draw a front elevation.

   b Make a sketch of the solid.

**Now go back to the list of objectives at the start of this chapter.
How confident do you now feel about each of them?**

17 Angles

- Complete these questions to assess how much you remember about each topic. Then mark your work using the answers at the back of the book.

- If you score well on all sections, you can go straight to the Revision checklist and Exam-style questions at the end of the chapter. If you don't score well, go to the chapter section indicated and work through the examples and practice questions there.

1 For each of the following angles state whether it is acute, obtuse, reflex or a right angle.

a   **b**   **c**   **d**   **e**

Go to 17.1

2 Measure each angle.

a   **b**

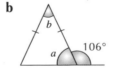

c   **d**

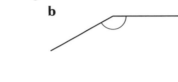

Go to 17.1

3 Work out the size of the angles marked with letters.   Go to 17.2, 17.3

a   **b**   **c**

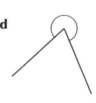

35°   106°   41°

4 Work out the size of the exterior angle of a 12-sided regular polygon.

Go to 17.4

5 For each diagram, find the bearing of B from A.   Go to 17.5

a   **b**

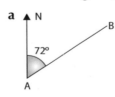   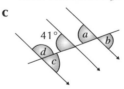

72°   152°

c   **d**

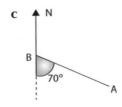   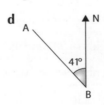

70°   41°

6 The diagram shows the position of three towns A, B and C. Find the bearing of:

    a A from B   **b** C from B.

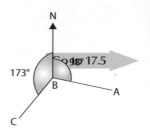

Go to 17.5

7 John walks due North for 100 metres.

He then walks on a bearing of 160° for 40 metres.

    a Draw a scale drawing to show his journey.

    b On what bearing, and how far, must John walk to return to his starting point?

Go to 17.6

17.1 Classifying angles

Angles are measured in units called **degrees**. The symbol is a small raised circle, °.

- An **acute** angle is greater than 0° and less than 90°.
- An **obtuse** angle greater than 90° and less than 180°.
- A **reflex** angle greater than 180° and less than 360°.
- A **right angle** is exactly 90°.

Acute angle   Obtuse angle   Reflex angle   Right angle

You measure the size of an angle using a protractor.

Make sure the cross is on the angle and that the base line on the protractor is on one arm of the angle. Read the angle from the scale that starts at 0°.

For this angle, read from the outer scale. The angle is 150°.

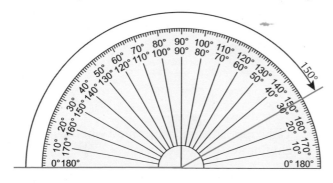

Exam tips

When measuring angles, check that you have read the angle from the correct scale. Always measure from 0°.

Practice questions

1 State whether each angle is acute or obtuse.

  a   **b**   **c**   **d**   **e**

2 Measure these angles.

a **b** **c** **d**

 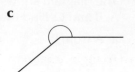 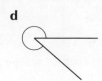

3 State whether each angle is acute, reflex, obtuse or a right angle.

a **b** **c** **d** **e**

   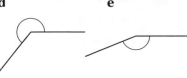

4 Look at this list of angles.

3°, 14°, 240°, 320°, 72°, 230°, 190°, 89°, 83°, 172°, 91°, 100°, 300°, 350°

List the angles which are:

a acute **b** obtuse **c** reflex.

5 Draw and label these angles.

a 70° **b** 120° **c** 220°

17.2 Angle facts

Whenever lines meet or intersect, the angles they form follow certain rules.

Angles on a straight line add up to 180°.

$a + b + c = 180°$

Angles in a triangle add up to 180°.

$a + b + c = 180°$

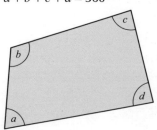

Angles in a quadrilateral add up to 360°.

$a + b + c + d = 360°$

Angles at a point add up to 360°.

$a + b + c + d = 360°$

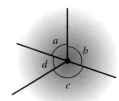

Vertically opposite angles (formed when two straight lines cross) are equal.

$a = b, c = d$

$a + d = b + c = 180°$

An exterior angle of a triangle is equal to the sum of the two opposite interior angles.

$c = a + b$

Since if

$a + b + d = 180°$ (angles in a triangle add up to 180°)

$d + c = 180°$ (angles on a straight line add up to 180°)

Then $c = a + b$

Labelling angles and triangles

When labelling a general triangle, the side opposite the vertex A is called a, the side opposite the vertex B is called b and the side opposite the vertex C is called c.

There are different ways of identifying an angle. When three letters are used, for example angle ABC, $\angle ABC$ or $A\hat{B}C$, the middle letter refers to the angle, i.e. the angle at B.

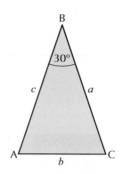

Example 1

Q Work out the size of the angles marked with letters.

a
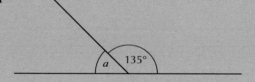

b
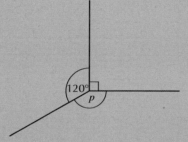

c
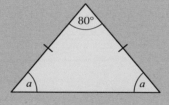

d
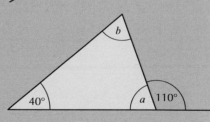

A
a $a + 135° = 180°$ (angles on a straight line sum to 180°)

$a = 180° - 135°$

$a = 45°$

b $p + 90° + 120° = 360°$ (angles at a point sum to 360°)

$p + 210° = 360°$

$p = 360° - 210°$

$p = 150°$

c $a + a + 80° = 180°$ (angles in a triangle sum to 180°)

$2a + 80° = 180°$

$2a = 180° - 80°$

$2a = 100°$

$a = 50°$

d $a + 110° = 180°$ (angles on a straight line sum to 180°)

$a = 70°$

$40° + b = 110°$ (the sum of two interior angles is equal to the opposite exterior angle of a triangle)

$b = 110° - 40°$

$b = 70°$

The angles opposite the equal sides in an isosceles triangle are equal.

Q Work out the sizes of the lettered angles in this diagram.

A $a = 64°$ (vertically opposite angles are equal)

$b = 180° - 64° = 116°$ (angles on a straight line add up to 180°)

$c = 64°$ (isosceles triangle; base angles are equal)

$d = 52°$ (angles in a triangle add up to 180°)

Exam tips Remember to give reasons for the angles you have calculated at each stage in your working. Always say that the angles **add up to** or **sum to** 180°/360° and not that they equal these values.

Practice questions

For questions 1 to 4, state the properties you have used in your calculations.

1 Work out the size of the angles marked with letters.

a

b

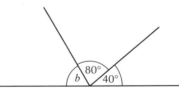

c

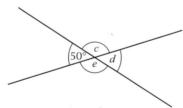

d

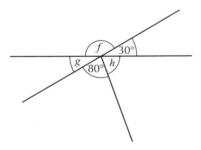

2 Work out the size of the angles marked with letters.

a

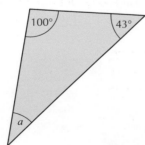

b

3 Work out the size of the angles marked with letters in these isosceles triangles.

a

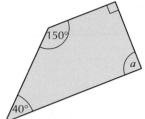

b

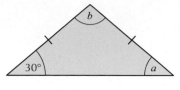

c

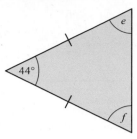

4 Work out the size of the angles marked with letters in these quadrilaterals.

a

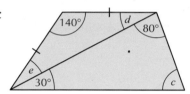

b

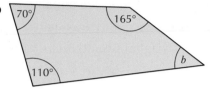

c

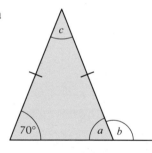

5 Work out the size of the angles marked with letters. Give reasons for each step in your calculations.

a

b

c

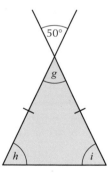

17.3 Angles in parallel lines

Three types of relationship are produced when a **transversal** crosses a pair of parallel lines.

- **Alternate** angles are equal.

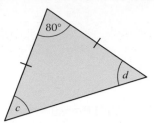

- **Corresponding** angles are equal.

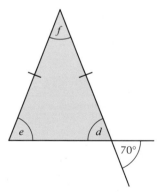

- **Allied** or **co-interior** angles add up to 180°.

$c + d = 180°$

You can use these relationships to find the value of some unknown angles.

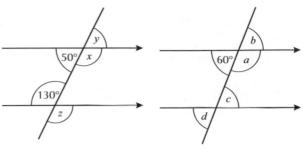

$x = 130°$ (alternate angles are equal)

$y = 50°$ (vertically opposite angles are equal)

$z = 130°$ (vertically opposite angles are equal)

$a = 120°$ (angles on a straight line add up to 180°)

$b = 60°$ (vertically opposite angles are equal)

$c = 60°$ (alternate angles are equal or corresponding angles to b are equal)

$d = 60°$ (vertically opposite angles (to c) are equal)

Example 3	**Q** BCD is a straight line. Explain why BE and CF must be parallel.

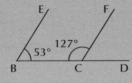

A The sum of the allied angles between parallel lines is 180°.

53° + 127° = 180°, so BE and CF are parallel.

Practice questions

1. Copy each of the following statements. Complete them using the correct 'angle relationship' from the box.

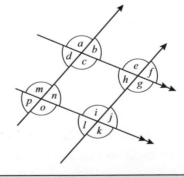

Angles a and c are...   Angles b and n are...

Angles b and h are...   Angles b and f are...

Angles g and i are...   Angles j and l are...

Angles p and d are...   Angles n and c are...

> vertically opposite angles   allied angles   corresponding angles   alternate angles

2 Work out the size of the angles marked with letters. Give reasons for your answers.

a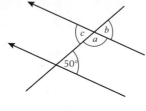

b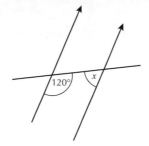

c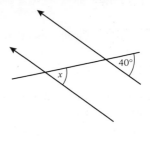

3 DC and GF are parallel.

Calculate angle FEH.

Give a reason for each step in your working out.

Take care to use correct angle notation, for example Angle ABC = 140°.

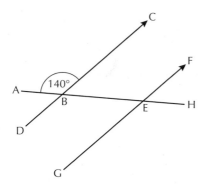

4 CB and FD are parallel and angle ABC = 75°.

Work out angle GED.

Give a reason for each step in your working.

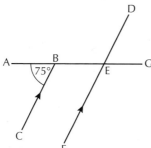

5 The lines AC, DF and GI are parallel.

Angle CBE = 45° and angle GHE = 125°.

Find the size of each of these angles.
Give a reason for each step in your working.

Angle EHI   Angle DEH   Angle BED   Angle BEH

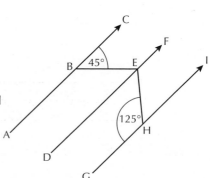

17.4 Angles in polygons

A regular polygon has all sides equal and all angles equal. (See Chapter 16.)

A polygon has **interior** and **exterior** angles. An exterior angle is formed when a side of the polygon is extended outside the shape.

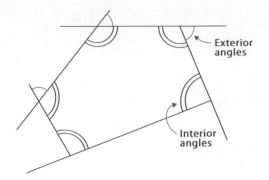

Exterior angles

Interior angles

For a polygon with n sides:

- Sum of exterior angles $= 360°$
- Interior angle + exterior angle $= 180°$
- Sum of interior angles $= (n - 2) \times 180°$ or $(2n - 4) \times 90°$
- For a regular polygon with n sides, exterior angle $= \dfrac{360°}{n}$

Example 4	**Q** Work out the size of an interior angle and an exterior angle of a regular hexagon.
	A A hexagon has 6 sides.
	Exterior angle of a regular hexagon is $\dfrac{360°}{6} = 60°$
	Interior angle of a regular hexagon is $180° - 60° = 120°$

Example 5	**Q** Find the sum of the interior angles of a regular pentagon.
	A A pentagon has 5 sides.
	Sum of interior angles of a regular pentagon is $= (n - 2) \times 180°$
	$= (5 - 2) \times 180°$
	$= 3 \times 180°$
	$= 540°$

Exam tips You could also work out the sum of the interior angles using this method:

One exterior angle is $\dfrac{360°}{5} = 72°$

One interior angle is $180° - 72° = 108°$

Sum of interior angles is $108° \times 5° = 540°$

Example 6	**Q** A regular polygon has an interior angle of $108°$. How many sides does it have?
	A Calculate the size of the exterior angle first.
	$180° - 108° = 72°$
	$360° \div 72° = 5$ sides

Example 7

Q Work out the size of angle y in this polygon.

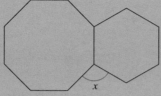

A Total sum of interior angles is $(6 - 2) \times 180° = 720°$

$$y + 126° + 83° + 145° + 138° + 79° = 720°$$
$$y + 571° = 720°$$
$$y = 720° - 571°$$
$$y = 149°$$

Example 8

Q The diagram shows a regular octagon and a regular hexagon.

Calculate the size of the angle marked x. Show all your working.

A Exterior angle of the hexagon $= \dfrac{360°}{6}$

$$= 60°$$

Interior angle of the hexagon $= 180° - 60°$

$$= 120°$$

Exterior angle of the octagon $= \dfrac{360°}{8}$

$$= 45°$$

Interior angle of the octagon $= 180° - 45°$

$$= 135°$$

$$x + 135° + 120° = 360° \quad \text{(angles at a point add up to } 360°\text{)}$$
$$x = 105°$$

Tessellations

A **tessellation** is a pattern of 2D shapes that fit together without leaving any gaps.

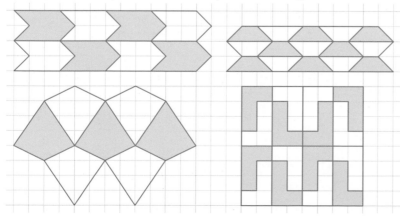

In order for shapes to tessellate, the angles at each point must add up to $360°$.

Regular pentagons do not tessellate. Each interior angle is $108°$, and $3 \times 108° = 324°$, so a gap of $360° - 324° = 36°$ is left.

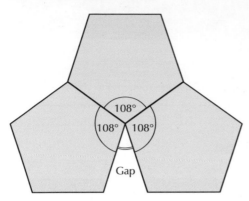

Practice questions

1 An irregular pentagon is divided into triangles as shown.

  a Use this to explain why the sum of the interior angles in a pentagon is $540°$.

  b Work out the size of each interior angle in a regular pentagon.

2 For each pentagon, work out the size of angle x.

a
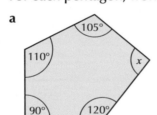

b
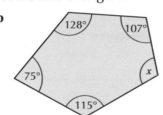

c
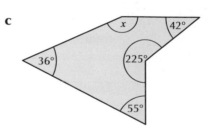

3 Find the size of angle a in each polygon.

a
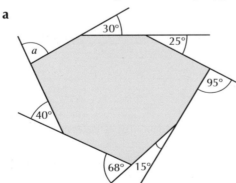

b
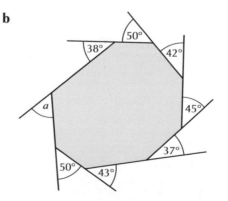

c
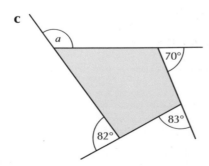

4 The diagram shows part of a regular polygon.

45°

The exterior angle is 45°.

How many sides does the polygon have?

5 The interior angles of a regular polygon are each 160°.

Explain why the polygon must have 18 sides.

6 A carpenter is making a seat in the shape of a regular nonagon (nine-sided shape) to fit around a post.

The diagram shows a plan view of the seat so far. To make the seat, the carpenter must cut the wood at the angle marked *a* on the diagram. What is the size of angle *a*?

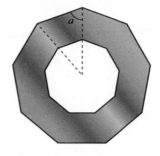

17.5 Bearings

The diagram shows the points of the compass.

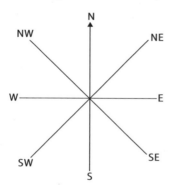

Directions can also be given as **bearings**. Bearings are used by aeroplanes and ships so they can travel in the right direction and avoid collisions.

Bearings are:

- always measured from the North (N).
- measured in a **clockwise** direction.
- written using three figures.

Bearings are also often used in scale drawing questions.

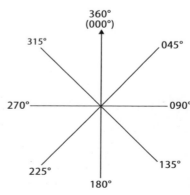

 Exam tips To find the bearing of A from B, point North from B and work out the number of degrees you need to turn through in a clockwise direction to point in the direction of A.

Example 9

Q Find the bearing of P **from** Q in each diagram.

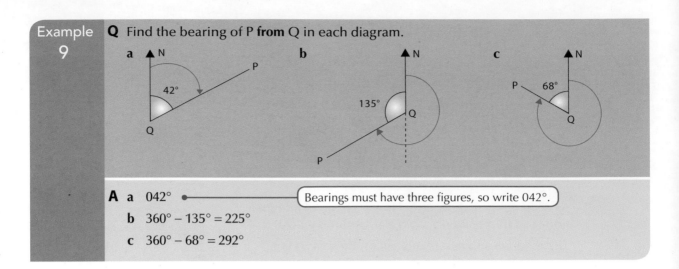

a N
42°
P
Q

b N
135°
Q
P

c N
P 68°
Q

A a 042° ●────────────── Bearings must have three figures, so write 042°.

b 360° − 135° = 225°

c 360° − 68° = 292°

In some bearings questions, it may be necessary to use more than one North line. As these will be parallel, you can then use the angle properties for parallel lines.

Example 10

Q Find the bearing of Q from P in each diagram.

a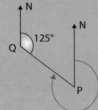
N
Q 125°
N
P

b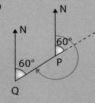
N
N 60°
60° P
Q

A a The North lines are parallel and allied angles add up to 180°.

So acute angle at NPQ is 180° − 125° = 55°.

Bearing of Q from P is 360° − 55° = 305°.

b 60° + 180° = 240° (using corresponding angles)

Example 11

Q A ship sails on a bearing of 143° to a buoy at A.

Work out the bearing on which the ship should sail to return to its starting point from the buoy.

A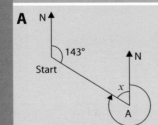
N
143°
Start
N
x
A

$x = 180° − 143°$ (allied angles add up to 180°)

$x = 37°$

Bearing is $360° − 37° = 323°$ (angles at a point add up to 360°)

Exam tips Drawing a sketch can help you to work out which angle is required.
Remember to write down your reasons at each stage of your working.

1 A boat sets out from a port in the direction SW.
After half an hour it changes direction by 90° anticlockwise.
On what bearing is the boat now sailing?

2 For each part, write down the bearing of P from Q.

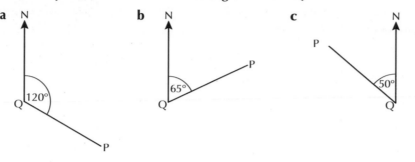

a   b   c   d

3 Each of these represents the bearing of P from Q. Draw a sketch for each one.
a 080°   b 110°   c 220°   d 340°

4 For each diagram, work out the bearing of Q from P.

a 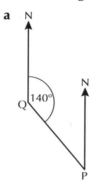   b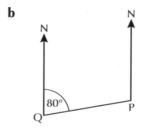

c 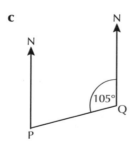   d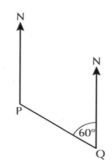

5 The bearing of a ship from a lighthouse is 025°.
What is the bearing of the lighthouse from the ship?

Hint

Sketch a diagram.

17.6 Scale drawings

Scale drawings are very useful for finding lengths that cannot be measured directly. When you draw scale diagrams, the lengths should be accurate to within 2 mm and the angles to within 2°.

Example
12

Q A ship sails from a harbour for 15 km on a bearing of 040°. It then continues due east for 20 km.

    a Make a scale drawing of this journey using a scale of 1 cm to 5 km.

    b How far will the ship have to sail to return to the harbour by the shortest route?

    c On what bearing will the ship have to sail to return to the harbour by the shortest route?

A **a** This diagram is not drawn to scale but is used to show what your diagram should look like.

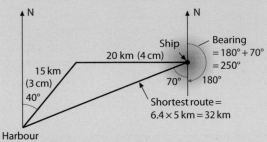

  b Shortest route = 6.4 × 5 km

                   = 32 km

  c Bearing = 70° + 180°

               = 250°

See Chapter 14 for more on scale drawings.

Practice questions

1 A boat sails due west for 8 km from port P to point A. It then sails on a bearing of 060° for 12 km.

    a Make a scale drawing for this journey.

    b How far and on what bearing will the boat have to sail to return to the port by the shortest route?

2 **a** Make an accurate scale drawing of this shape.

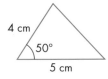

    b Measure the length of the third side.

3 Two boats set off from a point P at the same time.

One boat travels on a bearing of 040° at a speed of 10 mph. The other boat travels on a bearing of 160° at a speed of 15 mph.

By making a scale drawing, find the distance between the two boats after 1 hour.

4 Alan and Bob stand 1 km apart on a cliff top.

Bob is directly east of Alan.

From his position, Alan sees a boat on a bearing of 035°.
Bob sees the same boat on a bearing of 340°.

By making a scale drawing with a scale of 10 cm : 1 km,
find out how much closer Bob is to the boat than Alan.

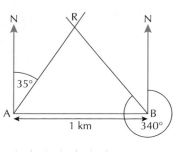

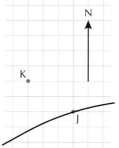

5 The diagram shows the positions of a jetty, J, and a boat K.

Copy the diagram onto centimetre-squared paper.

a What is the bearing of K from J?

A second boat, H, is due north of J
and on a bearing of 080° from K.

b Mark the position of H on your scale
drawing.

REVISION CHECKLIST

- Angles on a straight line add to 180°.
- Angles around a point add to 360°.
- The sum of angles in a triangle is 180°.
- The sum of angles in a quadrilateral is 360°.
- Alternate and corresponding angles are equal.
- Allied or co-interior angles add up to 180°.
- Polygons have interior and exterior angles.
- For a polygon with n sides:
  - Sum of exterior angles = 360°
  - Interior angle + exterior angle = 180°
  - Sum of interior angles = $(n - 2) \times 180°$ or $(2n - 4) \times 90°$
- For a regular polygon with n sides, an exterior angle = $\dfrac{360°}{n}$.
- Three-figure bearings are always measured from the North in a clockwise direction.

Exam-style questions

1 Make a scale drawing of the tower and work out its
height.

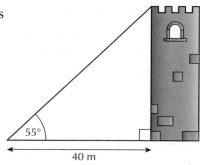

55°

40 m

2 Sita is facing North.

She makes a $\frac{3}{4}$-turn anticlockwise, followed by a clockwise turn of 135°.

In which direction is she now facing?

3 ABC is a straight line.

Work out the size of angle x.

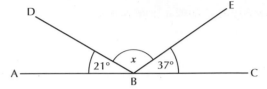

4 ABC is a right-angled triangle.

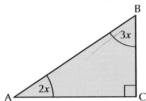

Work out the value of x.

5 Triangle ABC is isosceles.

Angle ACB = 74°

Find the size of angle BAC.

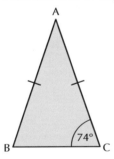

6 ABCD is a quadrilateral and E is a point on CD.

Angle ABE = angle ADE

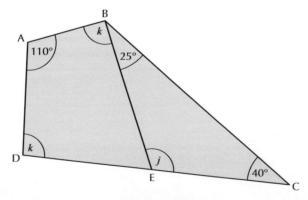

  a Work out the size of angle j. Give a reason for your answer.

  b Work out the size of angle k. Give a reason for your answer.

7 ABC is a triangle and C is a point on line AD.

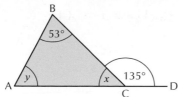

Work out the values of x and y. Give reasons for your answers.

8 The angles of a kite are as shown.
Calculate the value of x.

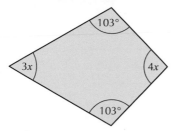

9 PQRS is a quadrilateral.

QPR and PRS are isosceles triangles.

QPS = 100° and angle PSR = 80°.

Calculate the size of angle PQR.

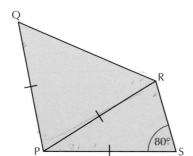

10 ABC is a triangle.

ABD is an equilateral triangle.

Work out the size of the angle marked x.

Give reasons for each stage of your working.

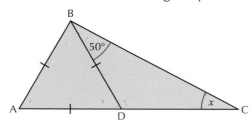

11 Kyle says, 'The bearing from A to B is 032°.'

Explain why Kyle is wrong.

12 Make an accurate drawing of this diagram.

Measure the length of QR.

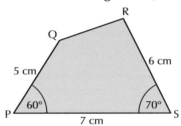

13 Work out the value of x. Give reasons for your answer.

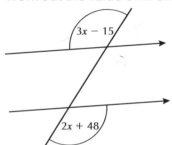

14 ABCD is a quadrilateral.

Is AB parallel to DC?

Explain your answer.

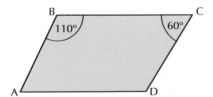

15 Three boats depart from a jetty at the same time and at the same speed of 12 miles per hour.

A sails on a bearing of 020°.

B sails on a bearing of 085°.

C sails on a bearing of 140°.

By making a scale drawing of the positions of the boats after 1 hour, decide whether boat A or boat C is closer to boat B and by how much.

16 AC and DG are parallel lines.

Angle BFG = 130°

Angle ABE = 80°

Give reasons why triangle BEF must be isosceles.

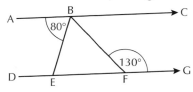

17 AB and CD are parallel lines.

Is EF parallel to CD and AB?

Give reasons for your answer.

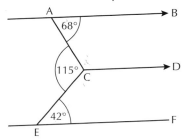

18 ABCD is a parallelogram.

Angle DAB = 140°, angle PDC = 25° and angle DPC = 75°.

Work out the sizes of the angles marked x and y, giving reasons for your answer.

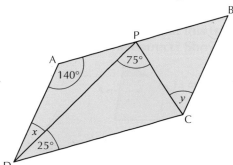

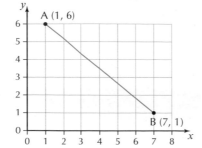

4 Calculate the length of the line segment AB.
Leave your answer in surd form.

Go to 18.2

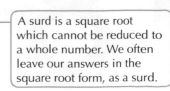

A surd is a square root which cannot be reduced to a whole number. We often leave our answers in the square root form, as a surd.

5 Calculate the height of this isosceles triangle. Give your answer in metres, correct to three significant figures.

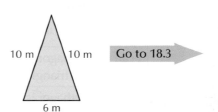

Go to 18.3

18.1 Calculating an unknown length

In a right-angled triangle, the side opposite the right angle is called the hypotenuse. The hypotenuse is the longest side in the triangle as it is opposite the largest angle. Pythagoras' theorem states that:

'For any right-angled triangle, the square on the hypotenuse is equal to the sum of the squares on the other two sides.'

You can use this to work out the length of the third side from any two known sides.

In this triangle $c^2 = a^2 + b^2$

Calculating the length of the hypotenuse

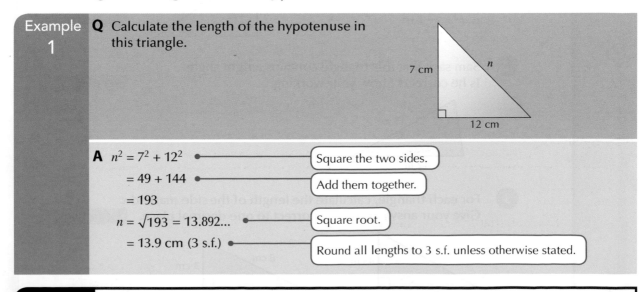

Example 1

Q Calculate the length of the hypotenuse in this triangle.

7 cm n 12 cm

A $n^2 = 7^2 + 12^2$ — Square the two sides.

$= 49 + 144$ — Add them together.

$= 193$

$n = \sqrt{193} = 13.892...$ — Square root.

$= 13.9$ cm (3 s.f.) — Round all lengths to 3 s.f. unless otherwise stated.

Exam tips Make sure you write down at least five significant figures before rounding.

Exam tips A surd is a square root whose value cannot be worked out exactly. You leave your answer in surd form when you are asked to give an exact answer. For the example above, this means $\sqrt{193}$ cm.

Calculating the length of a shorter side

Example 2

Q Calculate the length of p.

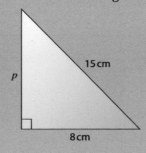

A
$$15^2 = p^2 + 8^2$$
$$15^2 - 8^2 = p^2$$
$$225 - 64 = p^2$$
$$161 = p^2$$
$$\sqrt{161} = p$$
$$p = 12.7 \text{ cm (3 s.f.)}$$

Remember to subtract when calculating a shorter length.

Exam tips When finding the length of the longest side: Square, Square, ADD, Square root.
When finding the length of one of the shorter sides: Square, Square, SUBTRACT, Square root.

Example 3

Q Molly says: 'Angle x is 90°.'
Explain how Molly knows this without measuring the angle.

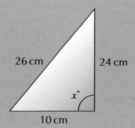

A $26^2 = 24^2 + 10^2$
$$676 = 576 + 100$$

The lengths obey Pythagoras' theorem, so the triangle must be right-angled.

Exam tips Do not round the numbers in your calculations before taking the square root.

Practice questions

 For each triangle, calculate the length x.

a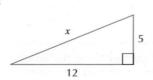

b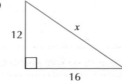

c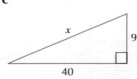

2 For each triangle, calculate the length x.

Give your answers correct to one decimal place.

a

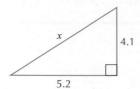

x 4.1 5.2

b

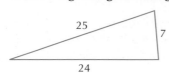

3.8 x 6

c

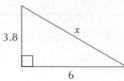

7 7 x

3 Is this a right angled triangle? You must explain your answer.

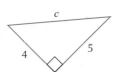

25 7 24

4 For each triangle, calculate the length c.

Give your answers correct to one decimal place.

a

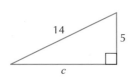

14 5 c

b

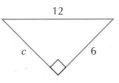

12 c 6

c

49.5 10 c

5 For each triangle, calculate the length c.

Give your answers as exact values.

a

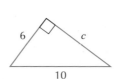

c 4 5

b

6 c 10

c

10 12.5 c

18.2 Calculating the length of a line segment on a coordinate grid

Example 4

Q Calculate the length of AB.

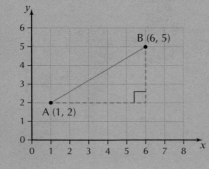

B (6, 5)

A (1, 2)

A Horizontal distance is $6 - 1 = 5$ ●——— Subtracting x coordinates

Vertical distance is $5 - 2 = 3$ ● Subtracting y coordinates

$$AB^2 = 5^2 + 3^2$$
$$= 25 + 9$$
$$= 34$$
$$AB = \sqrt{34}$$
$$= 5.83 \text{ units (3 s.f.)}$$

Exam tips | Unless you are told to give your answer correct to a certain number of decimal places or significant figures, you can leave your answer as a surd. Remember to include the units.

Practice questions

1 Calculate the length of the line segment PQ.

Give your answer correct to one decimal place.

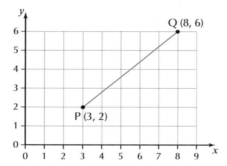

2 Calculate the length of the line segment XY.

Give your answer correct to one decimal place.

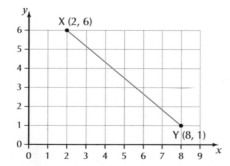

3 Calculate the distance between each of the pairs of points.

Give your answers correct to three significant figures.

a (2, 5) and (5, 2) **b** (4, 0) and (9, 7) **c** (0, 8) and (8, 0)

4 Calculate the length of the line segment AB.

Give your answer correct to three significant figures.

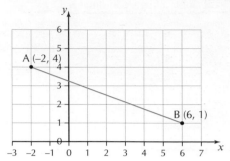

5 Calculate the distance between these pairs of points.

Give your answers correct to three significant figures.

a (–3, 2) and (4, 5)   **b** (2, 3) and (8, –3)   **c** (–3, –2) and (4, –3)

18.3 Solving problems

Example 5

Q Calculate the vertical height of this isosceles triangle.

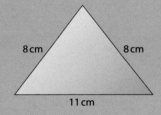

8 cm   8 cm

11 cm

A In an isosceles triangle the line of symmetry from the vertex to the base forms two congruent right-angled triangles. Draw a perpendicular line from the apex to the base line to obtain a right-angled triangle.

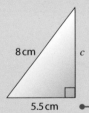

8 cm c

5.5 cm   The line of symmetry bisects the base.

Using Pythagoras' theorem gives:

$8^2 = c^2 + 5.5^2$

$64 = c^2 + 30.25$

$64 - 30.25 = c^2$

$33.75 = c^2$

$\sqrt{33.75} = c$

$c = 5.8094...$

$c = 5.81$ cm (3 s.f.)

Remember to write down the answer from your calculator to 5 sf before rounding to three significant figures.

Example 6

Q The diagram shows the floor plan of room. Karl wants to put beading around the edges of the room. He will leave a 1 m gap for the door. He can buy beading in lengths of 2.5 metres. Each length costs £1.74.

Work out the total cost of the beading for the room.

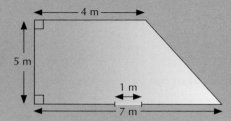

A $\sqrt{(5^2 + 3^2)} = 5.83\ldots$ ●————— Use Pythagoras' theorem to calculate the length of the unknown side.

Total length needed is
$5 + 4 + 7 + 5.83\ldots - 1 = 20.83\ldots$ m.

The number of lengths needed is $20.83\ldots \div 2.5 = 8.332$

You can only purchase whole lengths of beading. 8 lengths will not be enough, so you will need 9 lengths.

$9 \times £1.74 = £15.66$

Exam tips Check that you have actually answered the question that was asked and that your answer makes sense in context.

Practice questions

1 Calculate the length marked d.

Give your answer correct to three significant figures.

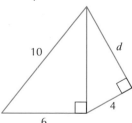

2 The diagram shows the plan of a piece of land. Bob is going to make a raised vegetable plot by placing concrete blocks around the edges of the piece of land.

Each concrete block is 40 cm long and costs £1.50.

How much will it cost Bob to buy the concrete blocks for the vegetable plot?

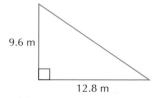

3 Calculate the length marked h.

Give your answer correct to three significant figures.

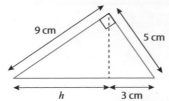

4 A ladder, 6 m long, leans against a wall.

The ladder reaches 5 m up the wall.

How far from the base of the wall is the foot of the ladder?

Give your answer correct to three significant figures.

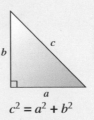

5 The diagram shows one end of a shed.

What is the height of the shed?

Give your answer correct to the nearest centimetre.

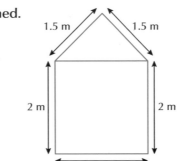

REVISION CHECKLIST

- Pythagoras' theorem states that:
  - 'For any right-angled triangle, the square on the hypotenuse is equal to the sum of the squares on the other two sides.'

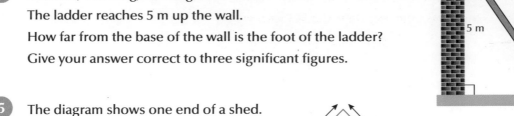

$$c^2 = a^2 + b^2$$

- Pythagoras' theorem can be used to find the length of the third side of a right-angled triangle when the lengths of the other two sides are known. Pythagoras' Theorem can be used to prove that an angle is a right angle.

Exam-style questions

1 Work out the length of x.

Give your answer correct to one decimal place.

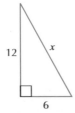

2 ABC is a right-angled triangle.

Calculate the length of AB.

Give your answer correct to three significant figures.

3 A ladder leans against a wall as shown.

How far up the wall does the ladder reach?

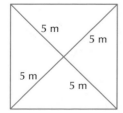

4 Which of these are true for this triangle? Write down all that apply.

$$x^2 = y^2 + z^2 \qquad y^2 = x^2 + z^2 \qquad z^2 = x^2 + y^2$$

$$x = \sqrt{y^2 - z^2} \qquad z = \sqrt{y^2 - x^2} \qquad y = \sqrt{z^2 - x^2}$$

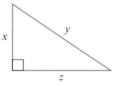

5 The diagram shows a patio in the shape of a trapezium.

How many edging pieces, each 50 cm long, are needed to go around all four edges of the patio?

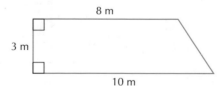

6 Calculate the perimeter of the square.

Give your answer correct to the nearest centimetre.

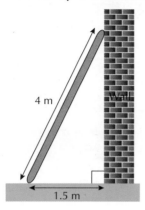

7 An 80 cm rod rests in a cylindrical bin with base diameter 24 cm and height 40 cm.

What length of the rod sticks out of the bin?

Give your answer correct to one decimal place.

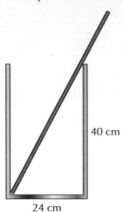

40 cm

24 cm

8 The diagram shows a shape made from a square and an isosceles triangle.

Calculate the perimeter of the shape.

Give your answer correct to one decimal place.

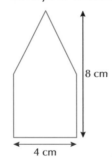

8 cm

4 cm

> **Hint**
>
> Make a copy of the diagram and add some lines to help you work out the height of the isosceles triangle.

9 The diagram shows a cake made in a square pan with a corner sliced off.

A ribbon is to be placed around the outside edges of cake.

Work out the total length of the ribbon needed.

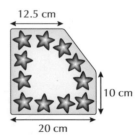

12.5 cm

10 cm

20 cm

10 Points A and B are plotted on a coordinate grid. B has coordinates (5, 8) and the midpoint of AB has coordinates (2, 3).

Work out the length of AB.

Give your answer correct to one decimal place.

11 A rhombus has diagonals of length 8 cm and 10 cm.

Work out the perimeter of the rhombus.

Give your answer correct to one decimal place.

12 The diagonals of rectangle ABCD meet at the point M, as shown in the diagram.

Calculate the length of AM.

Give your answer correct to one decimal place.

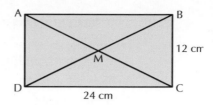

13 Kim is flying a kite.

The length of the string is 60 m.

Amber stands directly below the kite and 40 m from Kim.

Kim holds the end of the string 1.4 m above the ground.

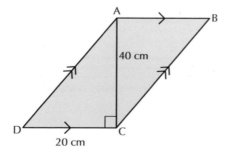

How high is the kite above the ground?

Give your answer in metres, correct to two decimal places.

14 The diagram shows a parallelogram, ABCD.

Calculate the perimeter of the parallelogram.

Give your answer correct to one decimal place.

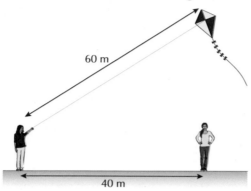

15 The diagram shows a rectangle ABCD.

E is a point on CD.

The diagonal BD is $\sqrt{52}$ cm, BE is 5 cm and CE is 3 cm.

Work out the length of the CD.

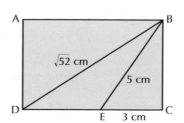

16 The diagram shows a square with diagonal of length d, and an equilateral triangle with height h.

The perimeter of the triangle is equal to the perimeter of the square.

Which is the longer length, d or h?

You must explain your answer.

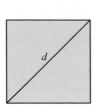 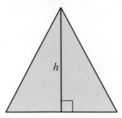

> ### Hint
>
> Try a value for the side of the square – a multiple of 3 units works well because you need to divide the perimeter by 3 to find the length of one side of the triangle. For a general algebraic solution, let one side of the square be $3x$.

17 ABC are points with coordinates (1, 7), (5, 6) and (6, 2) respectively.

Show that triangle ABC is isosceles.

18 Points A(3, 4) and C(9, 6) are diagonally opposite vertices of a square.

Work out the side length of the square.

Give your answer correct to two decimal places.

19 A, B and C have coordinates (3, 2), (2, 9) and (6, 5) respectively.

Is ABC a right-angled triangle?

Show your working.

20 W, X, Y and Z are points with coordinates (4, 6), (5, 10), (9, 9) and (8, 5) respectively.

Show that the quadrilateral WXYZ is a square.

Now go back to the list of objectives at the start of this chapter.

How confident do you now feel about each of them?

19 Trigonometry

Objectives

Before you start this chapter, mark how confident you feel about each of the statements below:

	▶	▶▶	▶▶▶
I know and can apply the trigonometric ratios to find an unknown length in a right-angled triangle.			
I know and can apply the trigonometric ratios to find an unknown angle in a right-angled triangle.			
I know the exact values of $\sin\theta$, $\cos\theta$ and $\tan\theta$ for $\theta = 0°$, $30°$, $45°$, $60°$ and $90°$.			
I can use trigonometry to solve 2D problems, including angles of elevation and depression.			
Stretch objectives Go to www.collins.co.uk/edexcelpost16			
I can apply Pythagoras' theorem, bearings and trigonometry to more complex problems.			

Check-in questions

- Complete these questions to assess how much you remember about each topic. Then mark your work using the answers at the back of the book.

- If you score well on all sections, you can go straight to the Revision checklist and Exam-style questions at the end of the chapter. If you don't score well, go to the chapter section indicated and work through the examples and practice questions there.

1   Copy and complete the statements by writing values in the blank spaces using the values from the box on the right.

Go to 19.1

$\sin \underline{\quad} = \dfrac{\sqrt{3}}{2}$ $\sin 45° = \underline{\quad}$

$\cos \underline{\quad} = \dfrac{\sqrt{3}}{2}$ $\cos \underline{\quad} = 1$

$\cos 90° = \underline{\quad}$ $\tan \underline{\quad} = 1$

$\tan 60° = \underline{\quad}$ $\tan 30° = \underline{\quad}$

60	0
$\sqrt{3}$	1
$\dfrac{\sqrt{3}}{3}$	30
$\dfrac{\sqrt{2}}{2}$	45

2   For each diagram, work out the value of x. Give your answers correct to two decimal places.

Go to 19.2

a

b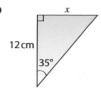

3 For each diagram, work out the size of the angle labelled x.
Give your answers correct to one decimal place.
Go to 19.3

a

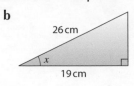

12 cm
x
15 cm

b

26 cm
x
19 cm

4 The diagram represents a vertical mast, PN, supported by two metal cables, PA and PB, fixed to the horizontal ground at A and B.

BN = 12.6 m

PN = 19.7 m

angle PAN = 48°

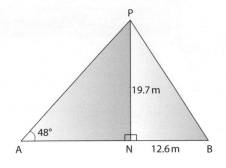

P
19.7 m
48°
A   N   12.6 m   B

a Calculate the size of angle PBN. Give your answer correct to three significant figures.

b Calculate the length of the metal cable PA. Give your answer correct to three significant figures.
Go to 19.3, 19.4

5 As part of an outward bound course, three groups of college students set up their tents at points A, B and C.

A is 7.6 km due west of B.

C is 9.8 km due north of B.

Calculate the bearing and distance of tent A from tent C. Give your answers correct to three significant figures.
Go to 19.3

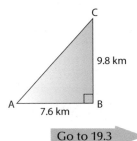

C
9.8 km
A   B
7.6 km

19.1 Trigonometric ratios

Trigonometry is a branch of mathematics that involves relationships between lengths and angles of triangles.

In a right-angled triangle, the sine, cosine and tangent ratios can be defined for an angle θ.

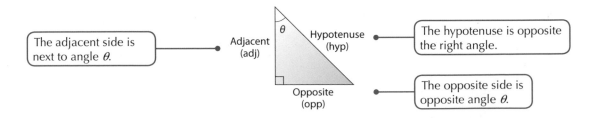

The adjacent side is next to angle θ.

Adjacent (adj)

θ

Hypotenuse (hyp)

The hypotenuse is opposite the right angle.

Opposite (opp)

The opposite side is opposite angle θ.

The trigonometric ratios are:

$$\text{Sine } \theta = \frac{\text{Opposite}}{\text{Hypotenuse}}$$

$$\text{Cosine } \theta = \frac{\text{Adjacent}}{\text{Hypotenuse}}$$

$$\text{Tangent } \theta = \frac{\text{Opposite}}{\text{Adjacent}}$$

You can use **SOH – CAH – TOA** to help you remember the ratios, for example, TOA means $\tan \theta = \frac{\text{opp}}{\text{adj}}$.

Exact values

The sine, cosine and tangent of some angles can be written exactly.

	0°	30°	45°	60°	90°
sin θ	0	$\frac{1}{2}$	$\frac{\sqrt{2}}{2}$	$\frac{\sqrt{3}}{2}$	1
cos θ	1	$\frac{\sqrt{3}}{2}$	$\frac{\sqrt{2}}{2}$	$\frac{1}{2}$	0
tan θ	0	$\frac{\sqrt{3}}{3}$	1	$\sqrt{3}$	undefined

Exam tips

You need to learn these values.

Example 1

Q Calculate the exact value of x.

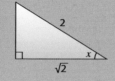

A First label the sides in relation to the angle.

You are given the lengths of the adj and hyp, so use cos.

$$\cos x = \frac{\text{adj}}{\text{hyp}}$$

$$= \frac{\sqrt{2}}{2}$$

$$x = 45°$$

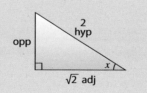

Practice questions

1. For each triangle, calculate the exact value of angle x.

a

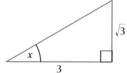

b

2 For each triangle, calculate the exact value of angle x.

a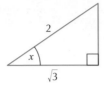

b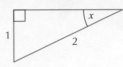

3 For each triangle, calculate the exact value of angle x.

a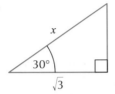

b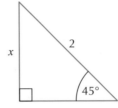

4 For each triangle, calculate the exact value of x.

a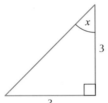

b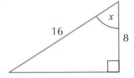

5 For each triangle, calculate the exact value of angle x.

a

b

c

19.2 Calculating a length

To calculate a length using trigonometric ratios:

• Label the sides in relation to the chosen angle.

• Decide which ratio you need to use.

• Substitute the required values.

When using your calculator, make sure that you are working in degrees. D or DEG will appear on the display.

The abbreviated forms, (sin, cos and tan) are used on calculator keys. To find the sine, cosine and tangent of an angle, press the appropriate key followed by the angle.

For sine 72°, press `sin` `7` `2` `=`

Example 2

Q Find the value of y.

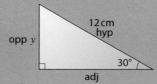

A $\sin 30° = \dfrac{\text{opp}}{\text{hyp}}$ You are given the hyp and need to find the opp, so use sin.

$\sin 30° = \dfrac{y}{12}$ Substitute the values.

$12 \times \sin 30° = y$ Multiply both sides by 12.

$y = 6$ cm

Exam tips It is good practice to write $12 \times \sin 30°$ and not $\sin 30° \times 12$ to ensure that you do not work out $\sin 360°$ by mistake.

Practice questions

1 For each triangle, calculate the length marked x.

a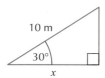

b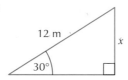

c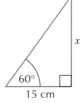

2 Use your calculator to work out:

a $\sin 35°$ **b** $\cos 80°$ **c** $\tan 25°$

3 For each triangle, calculate the length marked x.

a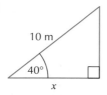

b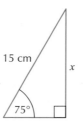

c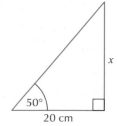

4 Julie is trying to find the value of x in triangle shown.

  a How can you tell from her answer that Julie has made a mistake?

  b What mistake has she made?

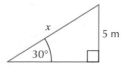

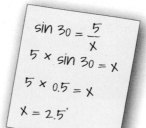

5 For each triangle, use the appropriate trigonometric ratio to calculate the value of x.

a

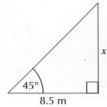

45°
8.5 m
x

b

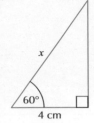

x
60°
4 cm

c

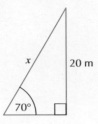

x
20 m
70°

19.3 Calculating an angle

To calculate an angle using trigonometric ratios:

• Label the sides in relation to the required angle.

• Decide which ratio you need to use.

• Substitute the required values.

• Calculate the inverse function.

The inverse function on your calculator may be labelled **SHIFT** **2ⁿᵈ F** or **INV**.

To find angle x when tan x = 0.75 press **SHIFT** **tan** **0** **·** **7** **5** **=** .

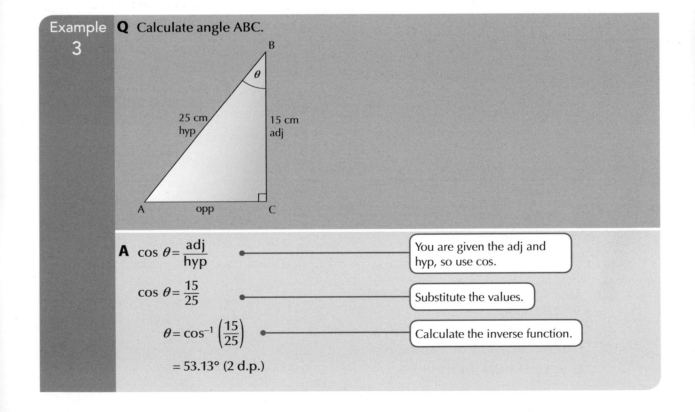

Example 3

Q Calculate angle ABC.

B
θ
25 cm hyp
15 cm adj
A
opp
C

A $\cos \theta = \dfrac{\text{adj}}{\text{hyp}}$ ———— You are given the adj and hyp, so use cos.

$\cos \theta = \dfrac{15}{25}$ ———— Substitute the values.

$\theta = \cos^{-1}\left(\dfrac{15}{25}\right)$ ———— Calculate the inverse function.

$= 53.13°$ (2 d.p.)

Example 4	**Q** As part of an orienteering award, a group of college students walk 37 km due north and then 42 km due east.

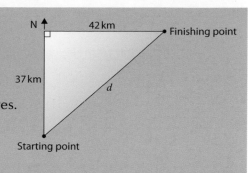

Q As part of an orienteering award, a group of college students walk 37 km due north and then 42 km due east.

a Calculate the direct distance between the starting point and the finishing point. Give your answer correct to two significant figures.

b Calculate the bearing of the students from their starting point. Give your answer correct to the nearest degree.

A a Use Pythagoras' theorem to find the distance (d).

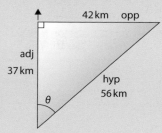

$$d^2 = 37^2 + 42^2$$
$$= 1369 + 1764$$
$$= 3133$$
$$d = \sqrt{3133}$$
$$= 55.973...$$
$$= 56 \text{ km (2 s.f.)}$$

b Use trigonometry to find the bearing.

$$\tan \theta = \frac{\text{opp}}{\text{adj}}$$
$$= \frac{42}{37}$$
$$\theta = \tan^{-1}(1.135...) \qquad \text{Do not round this value.}$$
$$= 48.621...° \qquad \text{Remember, bearings must be three figures.}$$

The bearing of the students from their starting point is 049° (to the nearest degree).

Exam tips   Be guided by the number of marks in a question. Questions with lots of marks usually involve several steps of working.
Draw a sketch if one is not provided.

Exam tips   Check that you know how to use your calculator to calculate trigonometric functions and their inverses. Check your calculator is set to work in degrees (not radians).

Angles of elevation and depression

The angle of elevation is the angle measured from the horizontal upwards.

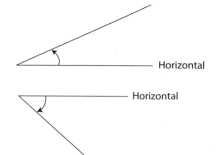

Horizontal

The angle of depression is the angle measured from the horizontal downwards.

Horizontal

Practice questions

1 For each triangle, calculate the size of angle a.

a

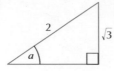

b

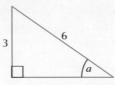

c

d

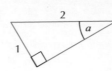

2 Use your calculator to work out the size of angle a in each case.

Give your answers correct to one decimal place.

a $\sin a = 0.8$ **b** $\sin a = 0.32$ **c** $\cos a = 0.25$

d $\cos a = 0.92$ **e** $\tan a = 0.45$ **f** $\tan a = 1.6$

3 For each triangle, calculate the size of angle a.

Give your answers correct to one decimal place.

a

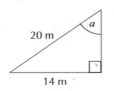

b

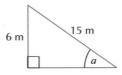

c

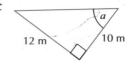

d

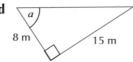

e

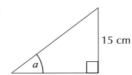

f

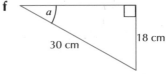

4 A vertical cliff is 80 metres high.

Richard looks at the top of the cliff from a point on the ground 200 metres from the base of the cliff.

Calculate the angle of elevation from Richard to the top of the cliff.

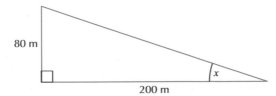

5 Sam is 20 metres from a tree.

He measures the angle of elevation to the top of the tree as 55°.

Calculate the height of the tree.

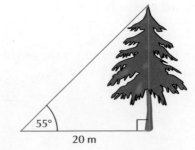

6 ABCD is a quadrilateral.

Angle BCD = angle BDA = 90°, angle A = 40°, BC = 8 cm, CD = 5 cm.

Calculate the length of AB.

Give your answer correct to three significant figures.

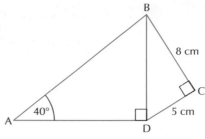

REVISION CHECKLIST

- Pythagoras' theorem: $c^2 = a^2 + b^2$

- Remember SOH – CAH – TOA:

$$\text{Sine } \theta = \frac{\text{Opposite}}{\text{Hypotenuse}} \qquad \text{Cosine } \theta = \frac{\text{Adjacent}}{\text{Hypotenuse}} \qquad \text{Tangent } \theta = \frac{\text{Opposite}}{\text{Adjacent}}$$

- You need to learn the sine, cosine and tangent of these angles can be written exactly.

	0°	30°	45°	60°	90°
$\sin \theta$	0	$\frac{1}{2}$	$\frac{\sqrt{2}}{2}$	$\frac{\sqrt{3}}{2}$	1
$\cos \theta$	1	$\frac{\sqrt{3}}{2}$	$\frac{\sqrt{2}}{2}$	$\frac{1}{2}$	0
$\tan \theta$	0	$\frac{\sqrt{3}}{3}$	1	$\sqrt{3}$	

- The angle of elevation is measured from the horizontal upwards.

- The angle of depression is measured from the horizontal downwards.

Exam-style questions

1 Which of the following statements is true?

  a $\tan 60° = 0$ **b** $\tan 60°$ is between 0 and 1

  c $\tan 60° = 1$ **d** $\tan 60°$ is greater than 1

2 Which of the following gives the length x in metres?

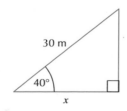

| $30 \times \cos 40°$ | $40 \times \cos 30°$ | $30 \times \sin 40°$ | $40 \times \sin 30°$ |

3 Calculate the size of angle *a*.

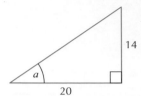

14

a

20

4 Calculate the length of LN.

Give your answer to an appropriate degree of accuracy.

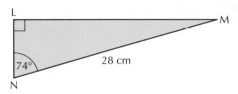

L

M

74°

28 cm

N

5 Calculate QR.

Give your answer to an appropriate degree of accuracy.

Q

7.25 m

24°

P

R

6 Calculate the size of angle *x*.

Give your answer correct to one decimal place.

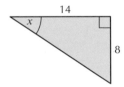

14

x

8

7 A tree 4 metres tall casts a shadow 10 metres long.

Calculate the angle of elevation of the sun.

Give your answer correct to one decimal place.

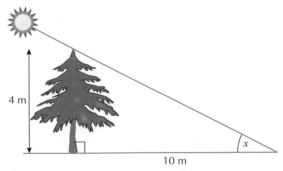

4 m

x

10 m

8   A boy is lying down on the top of a vertical cliff 60 metres high and looking out to sea.

He measures the angle of depression to a boat as 15°.

Work out the horizontal distance from the boat to the cliff.

9   Pat stands 40 metres from the base of a tree.

He measures the angle of elevation to the top of the tree as 32°.

Pat is 1.8 metres tall.

Calculate the height of the tree.

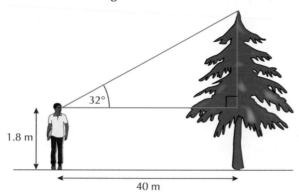

10 A boat sails 20 miles due north from a jetty.

It then sails 6 miles due west.

Calculate the bearing and distance of the jetty from the boat.

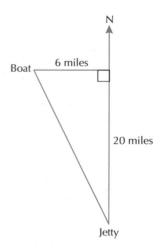

11 A tree is 24 metres east and 15 metres north from the corner of a building.

a Calculate the distance of the tree from the corner of the building.

b Work out the bearing of the tree from the corner of the building.

12 Work out the length of AB.

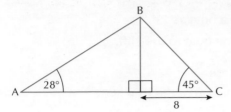

13 ABC is a triangle with area 50 cm².

Calculate the size of angle x.

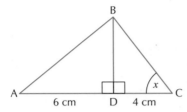

Now go back to the list of objectives at the start of this chapter.

How confident do you now feel about each of them?

20 Perimeter and area

Objectives

Before you start this chapter, mark how confident you feel about each of the statements below:

	▶	▶▶	▶▶▶
I can work out the perimeter of rectangles and triangles.			
I can work out the perimeter of parallelograms, trapezia and compound shapes.			
I can remember and use the formula for the area of a rectangle and the area of a triangle.			
I can remember and use the formula for the area of a trapezium.			
I can work out the area of a parallelogram.			
I can identify, name and draw parts of a circle, including *tangent*, *chord* and *segment*.			
I can recall and use the formula for the circumference of a circle and the area of a circle.			
I can understand and use π ≈ 3.142 and the π button on a calculator.			
I can work out the radius and diameter from the area or circumference of a circle.			
I can calculate the perimeter and area of composite shapes made from circles and parts of circles.			
I can calculate arc length, angles and areas of sectors of circles.			

Check-in questions

- Complete these questions to assess how much you remember about each topic. Then mark your work using the answers at the back of the book.
- If you score well on all sections, you can go straight to the Revision checklist and Exam-style questions at the end of the chapter. If you don't score well, go to the chapter section indicated and work through the examples and practice questions there.

1 Work out the area and perimeter of these shapes.
Remember to include the correct units.

Go to 20.1

a

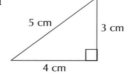

b

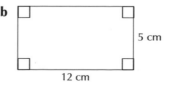

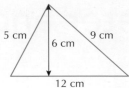

c

d

2 Work out the area of these shapes.

Go to 20.2

a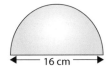

b

3 Calculate the circumference and area of each circle. Use π = 3.142.
Give your answers correct to two decimal places.

Go to 20.3

a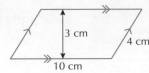

b

c

4 Work out the perimeter and area of this shape.
Use the π button.

Go to 20.3

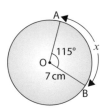

5 Calculate:

a the arc length (x)   b the area of the sector AOB.

Go to 20.4

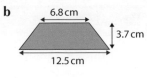

20.1 Plane shapes

The **area** of a 2D shape (or plane shape) is the amount of flat space that it covers. Common units of area are square millimetres (mm²), square centimetres (cm²) and square metres (m²).

(See Chapter 21 to convert between different units of area.)

The distance around the outside edge of a shape is called the **perimeter**. Remember to include the units with your answer.

1	2	3	4
5	6	7	8
9	10	11	12
13	14	15	16
17	18	19	

These make one whole square

This is half a square

Areas of irregular shapes can be found by counting the square units the shape covers.

This shape has an area of about 20.5 square units.

<table>
<tr><td>Example 1</td><td>Q Work out the perimeter of this shape.
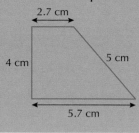</td><td>A Perimeter = 4 + 5 + 2.7 + 5.7
= 17.4 cm</td></tr>
</table>

Important formulae for area

You need to know these formulae.

Area of a rectangle

Area (*A*) = length (*l*) × width (*w*)

$A = l \times w$

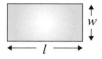

Area of a triangle

Area (*A*) = $\frac{1}{2}$ × base (*b*) × perpendicular height (*h*)

$A = \frac{1}{2} \times b \times h$

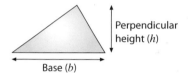

Area of a parallelogram

Area (*A*) = base (*b*) × perpendicular height (*h*)

$A = b \times h$

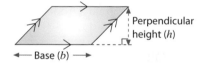

Area of a trapezium

Area = $\frac{1}{2}$ × sum of parallel sides × perpendicular distance between them

$A = \frac{1}{2} \times (a + b) \times h$

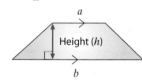

> **Hint**
>
> Remember the height must be at right angles to the base.

<table>
<tr><td>Example 2</td><td>Q Calculate the area of the parallelogram.
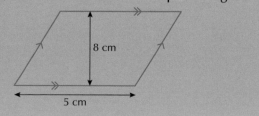</td><td>A $A = b \times h$
= 5 × 8
= 40 cm²</td></tr>
</table>

<table>
<tr><td>

Example 3

Q Calculate the area of this shape.

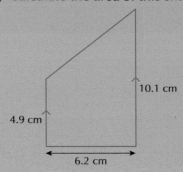

10.1 cm

4.9 cm

6.2 cm

</td><td>

A $A = \frac{1}{2} \times (a + b) \times h$

$= \frac{1}{2} \times (4.9 + 10.1) \times 6.2$

$= 46.5 \text{ cm}^2$

</td></tr>

<tr><td>

Example 4

Q The area of this triangle is 55 cm². Calculate its height correct to three significant figures.

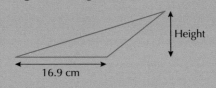

Height

16.9 cm

</td><td>

A $A = \frac{1}{2} \times b \times h$

$55 = \frac{1}{2} \times 16.9 \times h$ — Substitute values into the formula.

$55 = 8.45 \times h$

$h = \frac{55}{8.45}$ — Divide both sides by 8.45.

$= 6.5088...$

$h = 6.51 \text{ cm (3 s.f.)}$ — Round the final result only.

</td></tr>

<tr><td>

Example 5

Q The perimeter of this shape is 30 cm.

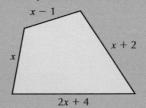

$x - 1$

$x + 2$

x

$2x + 4$

Work out the length of the longest side.

The perimeter of this shape is 42 cm.

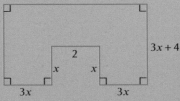

2

x x

$3x + 4$

$3x$ $3x$

Work out the value of x.

The triangle and the rectangle have equal perimeters.

(a) Write down and equation in x.

(b) Solve your equation to find x.

(c) Work out the perimeter of each shape.

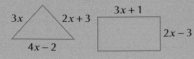

$3x$ $2x + 3$

$3x + 1$

$2x - 3$

$4x - 2$

</td><td>

A $x + x - 1 + 2x + 4 + x + 2 = 30$

$5x + 5 = 30$ — Set up an equation.

$5x = 30 - 5$ — Collect like terms.

$5x = 25$

$x = 25 \div 5$ — Subtract 5 from both sides.

$x = 5 \text{ cm}$ — Divide both sides by 5.

Length of the longest side is $2x + 4 = 2 \times 5 + 4$

$= 14 \text{ cm}$

$x = 1.5$

(a) $9x + 1 = 10x - 4$

(b) $x = 5$

(c) 46

</td></tr>
</table>

Practice questions

1 Calculate the area and perimeter of each shape.

a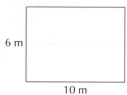

b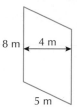

c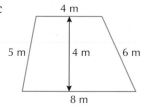

2 Calculate the area of each triangle.

a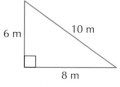

b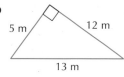

c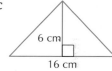

d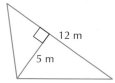

e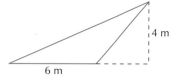

3 Which of these shapes has the largest area? You must explain your answer.

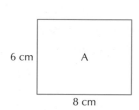

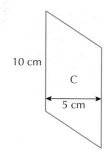

4 A triangle has a base of 14 cm and an area of 56 cm^2.
Work out the perpendicular height of the triangle.

5 An equilateral triangle has sides of length 12 cm.
A square has the same perimeter as the triangle.
Work out the area of the square.

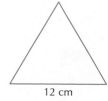

20.2 Compound shapes and problem solving

You can work out the area of more complex shapes by splitting them into simple shapes.

Example 6

Q The diagram shows the plan of a garden.

Lawn seed is sold in 500 g packets that cover 15 m². A packet of lawn seed costs £6.25. Work out the total cost of the lawn seed needed to cover the garden.

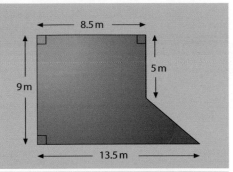

A Work out the area of the garden by splitting it into a rectangle and a triangle.

Area of rectangle = $l \times w$

$= 9 \times 8.5$

$= 76.5$ m²

Area of triangle = $\frac{1}{2} \times b \times h$

$= \frac{1}{2} \times 5 \times 4$

$= 10$ m²

Base of triangle is 13.5 − 8.5 = 5 m
Height of triangle is 9 − 5 = 4 m

Total area = 76.5 + 10

$= 86.5$ m²

Number of packets of lawn seed needed: $\frac{86.5}{15} = 5.77$ packets

Round up to next whole number ———● = 6 packets needed

Cost: £6.25 × 6 = £37.50

Example 7

Q The diagram shows the floor plan of a room. The cost of installing underfloor heating is £155 per square metre.

Work out the total cost of installing underfloor heating in this room.

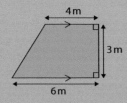

A Area of the floor $(A) = \frac{(a+b)}{2} \times h$

$= \frac{(4+6)}{2} \times 3$

$= 15$ m²

Total cost: 15 × 155 = £2325

Practice questions

1 Each of these shapes is made from a rectangle with a section removed.

Work out the area of each shape.

a

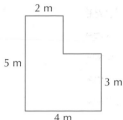

b

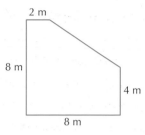

c

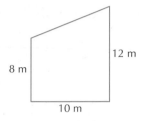

2 Work out the area of this shape.

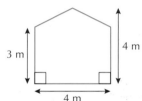

3 A garden is to be made from a lawn in the shape of a trapezium with a central flower bed in the shape of a square, as shown in the diagram.

The area shaded green is to be covered in grass seed. 50 g of grass seed is needed for each square metre of area. Grass seed costs £3 for each 500 g bag.

How much will it cost to cover the lawn with grass seed?

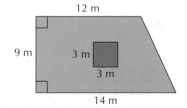

4 Mark wants to put fence panels along three sides of his garden as shown in the sketch.

The fence panels are each 2 metres long.

One post is needed at the each end of a panel.

Posts costs £4 and panels cost £12.

Work out the total cost of the fence panels and posts Mark needs for his garden.

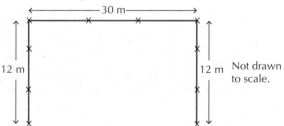

Not drawn to scale.

5 The diagram shows the floor plan of a playroom.

Wood flooring is sold in packs that cover an area of 1.5 square metres. Each pack costs £14.

What is the total cost of the wood flooring needed to cover the playroom floor?

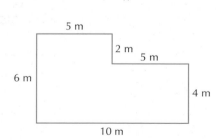

20.3 Circles

You need to be able to understand and use the vocabulary and formulae related to circles.

- The **circumference** of a circle is the distance around its outside edge.

- The **diameter** of a circle is a straight line that reaches from one side of a circle to the other, passing through the middle.

- The **radius** of a circle is the distance from the centre to any point on the circumference. It is half the diameter.

- Circumference = π × diameter

$\qquad\qquad\quad$ = 2 × π × radius

- Area of circle = π × (radius)^2

- A **tangent** is a line that touches the circle at one point only.

- An **arc** is part of the circumference of a circle.

- A **chord** is a line joining two points on the circumference that does not pass through the centre of the circle.

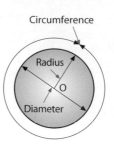

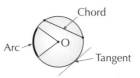

Exam tips The guidance on the front page of the calculator paper says: "If your calculator does not have a π button, take the value of π to be 3.142 unless the question instructs otherwise."

Example 8

Q Calculate $\quad$ **a** the perimeter and $\qquad$ **b** the area of this shape.

Use the π button on your calculator. Give your answers correct to three significant figures.

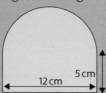

A **a** Length of straight sides: $\qquad$ 12 + 5 + 5 = 22 cm

$\qquad$ Circumference of semicircle: $\quad \dfrac{\pi \times d}{2} = \dfrac{\pi \times 12}{2}$

$\qquad\qquad\qquad\qquad\qquad\qquad\qquad\qquad\qquad$ = 18.849… cm

$\qquad$ Perimeter of shape: $\qquad$ 22 + 18.849… = 40.849…

$\qquad\qquad\qquad\qquad\qquad\qquad\qquad\qquad$ = 40.8 cm (3 s.f.)

$\quad$ **b** Area of rectangle: $\qquad l \times w = 12 \times 5$

$\qquad\qquad\qquad\qquad\qquad\qquad$ = 60 cm^2

$\qquad$ Area of semicircle: $\qquad \dfrac{\pi \times r^2}{2} = \dfrac{\pi \times 6^2}{2}$

$\qquad\qquad\qquad\qquad\qquad\qquad\qquad$ = 56.548… cm^2

$\qquad$ Total area: $\qquad$ 60 + 56.548… = 116.548…

$\qquad\qquad\qquad\qquad\qquad\qquad$ = 117 cm^2 (3 s.f.)

Example 9

Q A circle has an area of 60 cm². Calculate the radius of the circle correct to three significant figures. Use π = 3.142.

A $A = \pi \times r^2$

$60 = 3.142 \times r^2$ ⟵ Substitute the values into the formula.

$\dfrac{60}{3.142} = r^2$ ⟵ Divide both sides by 3.142.

$r^2 = 19.0961\ldots$

$r = \sqrt{19.0961\ldots}$ ⟵ Square root to find the value of r.

$= 4.37$ cm (3 s.f.)

Example 10

Q The diagram shows the plan of the inside of a running track.

The groundsperson is going to cover this area with grass seed. One sack of grass seed covers 275 m².

How many sacks of grass seed does the groundsperson need?

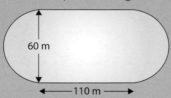

A Area of rectangle: $110 \times 60 = 6600$ m²

Area of circle: $\pi \times 30^2 = 2827.4$ m²

Total area: $6600 + 2827.4 = 9427.4$ m²

Number of sacks of grass seed required: $9427.4 \div 275 = 34.28\ldots$

He needs **35** sacks of grass seed.

Exam tips | Look at the context of the question to decide whether it is appropriate to round up or down.

Practice questions

1 Calculate the circumference of each circle.

Give your answers correct to one decimal place.

a 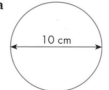 **b** 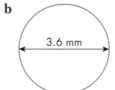 **c** 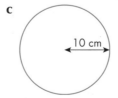 **d**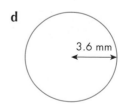

10 cm   3.6 mm   10 cm   3.6 mm

2 Calculate the area of each circle in Question 1.

Give your answers correct to one decimal place.

3 Calculate the area of this shape.

Give your answer correct to two decimal places.

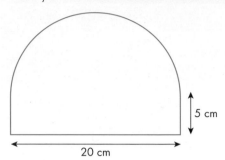

4 Calculate the shaded area in each diagram. Give your answers correct to two decimal places.

a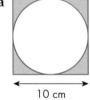

10 cm

b

20 cm

c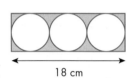

18 cm

5 **a** Each shape is made up of semicircles. Work out the shaded area.

Leave your answers in terms of π.

i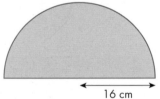

16 cm

ii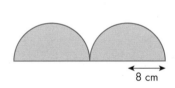

8 cm

iii

4 cm

b Using your answers to part **a i**, **ii** and **iii**, write down the total area of a shape made up of eight semicircles, each with a radius of 2 cm.

20.4 Arc length and sector area

Length of a circular arc

The length of a circular **arc** can be expressed as a fraction of the **circumference** of a circle.

Arc length = $\frac{\theta}{360°} \times 2\pi r$,

where θ is the angle subtended at the centre

A minor arc is less than a semicircle. A major arc is greater than a semicircle.

Example 11

Q Work out the length of the minor arc AB.

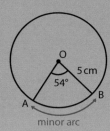
minor arc

A Arc length = $\dfrac{54°}{360°} \times 2 \times \pi \times 5$ ●————(O is the centre of the circle.)

$= \dfrac{3}{2}\pi$ cm (non-calculator paper) or 4.71 cm (2 d.p.) (calculator paper)

Area of a sector

The area of a **sector** can be expressed as a fraction of the area of a circle.

Area of a sector = $\dfrac{\theta}{360°} \times \pi r^2$, where θ is the angle subtended at the centre

A minor sector is less than a semicircle. A major sector is greater than a semicircle.

Example 12

Q Work out the area of the minor sector AOB.

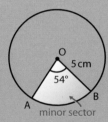
minor sector

A Area of sector = $\dfrac{54°}{360°} \times \pi \times 5^2$

$= \dfrac{15}{4}\pi$ cm² (non-calculator paper) or 11.78 cm² (2 d.p.) (calculator paper)

Exam tips | Learn the formulae carefully, especially those for the circumference and area of a circle.

Practice questions

1 Calculate the arc length of each sector correct to three significant figures.

a

60°
20 cm

b
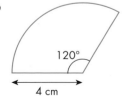
120°
4 cm

c
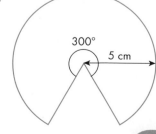
300°
5 cm

2 Calculate the area of each sector, giving your answers in terms of π.

a

60°

6 cm

b

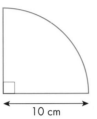

120°

9 cm

c

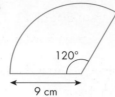

300°

2.4 m

3 Calculate the area and perimeter of this shape.
Give your answers correct to two decimal places.

10 cm

4 **a** Work out the shaded area in this shape.
Give your answer correct to two decimal places.

b Work out the perimeter of the shaded area
of the shape.
Give the answer in terms of π.

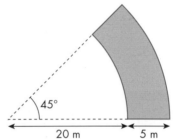

45°

20 m   5 m

5 A circle has an area of 25 cm². Calculate the radius of the circle.
Give your answer correct to two decimal places.

6 The diagram shows a running track.
Rachel is training for a race.
She wants to run at least 10 km each day.
What is the fewest number of laps of this
running track she needs to run every day?

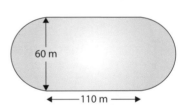

60 m

110 m

REVISION CHECKLIST

- Area is measured in square units.
- The formulae you need to learn to calculate the areas of 2D shapes are:
  - Area of a rectangle: $A = l \times w$
  - Area of a triangle: $A = \frac{1}{2} \times b \times h$
  - Area of a parallelogram: $A = b \times h$
  - Area of a trapezium: $A = \frac{1}{2}(a + b)h$
- The circumference of a circle is also its perimeter.
- Circumference of a circle: $C = \pi d$ or $2\pi r$
- Area of a circle: $A = \pi r^2$
- Arc length $= \frac{\theta}{360°} \times 2\pi r$
- Area of a sector $= \frac{\theta}{360°} \times \pi r^2$

Exam-style questions

1 Show each of these on a separate diagram.

a a diameter of a circle   **b** a chord of a circle   **c** a tangent of a circle

2 Nine identical rectangular pieces fit together as shown.
Work out the total area of the shape.

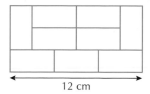

12 cm

3 Work out the area of this shape.

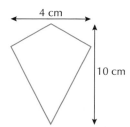

3 m

7 m

6 m

10 m

4 The total area of this shape is 32.25 m².
Work out the side length of the square.

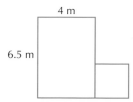

4 m

6.5 m

5 Work out the area of the kite shown in the diagram.

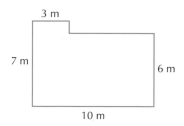

4 cm

10 cm

6 The diagram shows a rectangular lawn surrounded by a path. The path is 1 m wide.
Work out the area of the path.

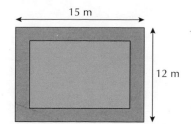

15 m

12 m

3.125

7 Pat is painting the doors in her house.

Each door measures 2.1 m by 1.1 m.

2.1 m

1.1 m

Pat wants to paint both sides of all of her seven doors.

A tin of paint costs £8. The guide on the tin says:

'1 tin covers 5 m²'

a Work out the total cost of painting all the doors.

b Pat finds that the first tin of paint covers less than 5 m². How might this affect your answer to part **a**?

8 The area of the triangle is five times greater than the area of the trapezium.

Work out the perpendicular height, h, of the trapezium.

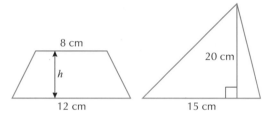

8 cm

h

12 cm

20 cm

15 cm

9 The diagram shows a circular mirror inside a frame in the shape of a trapezium.

Work out the area of the frame.

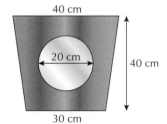

40 cm

20 cm

40 cm

30 cm

10 Work out the area of this semicircle.

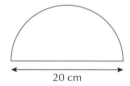

20 cm

11 The diagram shows a quarter circle.

Calculate its perimeter.

12 cm

12 Calculate the area of this shape.

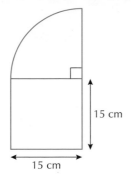

15 cm

15 cm

13 A circle and a square have the same area.

The radius of the circle is 10 cm.

Calculate the side length of the square.

Give your answer correct to one decimal place.

14 The diagram shows a semicircle and rectangle.

The semicircle and rectangle have the same area.

The radius of the semicircle is 15 cm.

Calculate the height, h, of the rectangle.

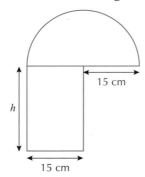

15 cm

h

15 cm

15 The shape in the diagram is made from a rectangle with two identical semicircles removed.

Work out the area of the shape.

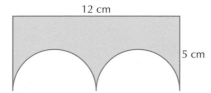

12 cm

5 cm

16 The semicircle and triangle shown have the same area.

Work out the value of h.

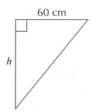

60 cm

60 cm

h

17 Eva bakes three identical circular cakes.

She has a ribbon exactly 1 m in length.

She puts ribbon round the circumference of each cake with no overlap and no gaps.

This uses all the ribbon.

Work out the diameter of each cake.

18 Twelve circles are cut from a rectangular sheet of metal, as shown.

Work out the shaded area.

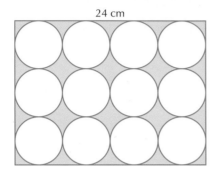

24 cm

19 Exactly half of this square is shaded.

Work out the radius of the circle.

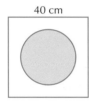

40 cm

20 Work out the shaded area of this shape.

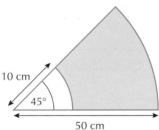

10 cm

45°

50 cm

**Now go back to the list of objectives at the start of this chapter.
How confident do you feel now about each of them?**

21 Surface area and volume

Objectives

Before you start this chapter, mark how confident you feel about each of the statements below:

	▶	▶▶	▶▶▶
I can recall and use the formulae for the surface area and volume of a cuboid.			
I can calculate the volume of a prism.			
I can calculate the surface area and volume of a cylinder.			
I can calculate the surface area and volume of pyramids and cones.			
I can calculate the surface area and volume of spheres and composite solids.			
I can convert between units of measure.			

Check-in questions

- Complete these questions to assess how much you remember about each topic. Then mark your work using the answers at the back of the book.
- If you score well on all sections, you can go straight to the Revision checklist and Exam-style questions at the end of the chapter. If you don't score well, go to the chapter section indicated and work through the examples and practice questions there.

1. Work out the volume of each prism.

   Give your answers correct to one decimal place.

   Use $\pi = 3.142$.

   Go to 21.1

   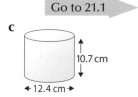

   a 2 cm, 8 cm, 6 cm **b** 7.2 cm, 12 cm, 9 cm **c** 10.7 cm, 12.4 cm

2. A cone with a vertical height of 2.1 m has a volume of 15 m^3.

   Calculate the radius of the base of the cone. Give your answer correct to three significant figures.

   Go to 21.2

3. How many cubic centimetres are there in 1 m^3?

   Go to 21.3

4. George has a piece of cloth that measures 0.5 m by 0.5 m.

   What is the area of the cloth:

   a in square metres **b** in square centimetres?

   Go to 21.3

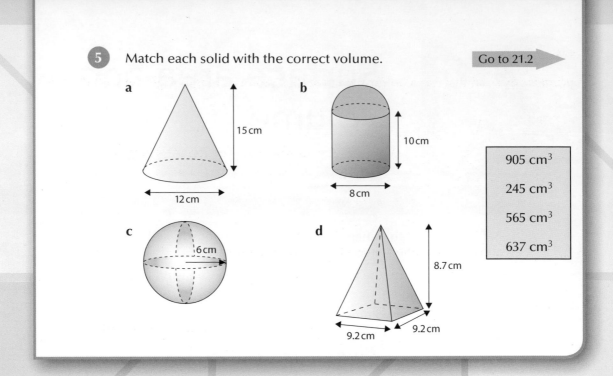

5 Match each solid with the correct volume.

Go to 21.2

21.1 Simple 3D shapes

The **volume** of a 3D shape is the amount of space it occupies. Commonly used units of volume include cubic millimetres (mm^3), cubic centimetres (cm^3) and cubic metres (m^3).

You can find the volume of some solid shapes by counting the number of centimetre cubes they are made from.

To calculate the surface area of a shape, work out the area of each face and then add them together. You need to learn the formulae for these shapes.

Cuboids

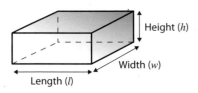

Volume (V) = length (l) × width (w) × height (h)

$V = lwh$

To calculate the surface area of a cuboid, work out the area of each face, then add together.

Surface area = $2hl + 2hw + 2lw$

Prisms

A **prism** is any solid that has a uniform **cross-section** along its length together with flat sides.

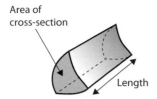

Area of cross-section

Length

Volume (*V*) = area of cross-section (*A*) × length (*l*)

Volume of a cylinder

A cylinder has two circular ends connected by a curved surface.

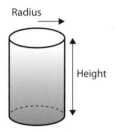

Radius

Height

Volume = area of cross-section × length/height

$V = \pi r^2 \times h$

Example 1

Q Calculate the volume of this prism.

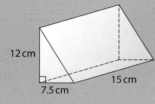

12 cm

7.5 cm

15 cm

A $V = A \times l$

$= (\frac{1}{2} \times b \times h) \times l$ — The area of the cross-section is the area of the triangle.

$= (\frac{1}{2} \times 7.5 \times 12) \times 15$

$= 675 \text{ cm}^3$ — Cubic units for volume

Exam tips   Remember to write down the units.

Practice questions

For questions 1–6, calculate the volume of each shape.

All lengths are in centimetres.

 1

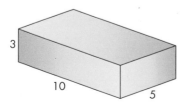

3

10

5

2

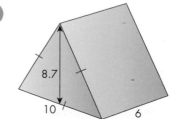

8.7

10

6

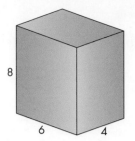

8

6   4

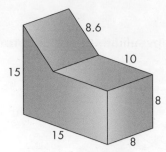

8.6

10

15

8

15   8

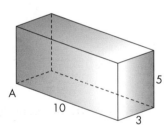

8

6

5

4

10

13

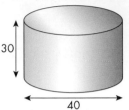

30

40

7 The diagram shows a cuboid A and a right-angled triangular prism B.

Show that the volume of B is twice the volume of A.

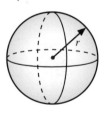

A   5

10   3

B   5   4   30

21.2 More complex solids

The formulae for these shapes are given on the formula page of the exam paper. In the exam, the relevant formulae for these shapes will be given with the question.

Spheres

Volume of a sphere $= \frac{4}{3}\pi r^3$

Surface area of a sphere $= 4\pi r^2$

Pyramids

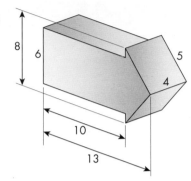

Volume of a pyramid $= \frac{1}{3} \times$ area of base $\times$ height

The total surface area of the pyramid can be found by drawing a net and then calculating the area of each face.

Cones

A cone is a pyramid with a circular base.

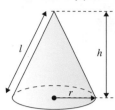

Volume of a cone $= \frac{1}{3} \times \pi \times r^2 \times$ height

$$V = \frac{1}{3} \pi r^2 h$$

Curved surface area of cone $= \pi rl$, where l is the slant height

Total surface area of a cone with a base $= \pi rl + \pi r^2$

Surface area of cylinders

The total **surface area** of a cylinder can be found by drawing a net.

Cylinder   Net of a cylinder

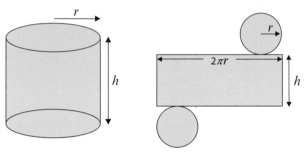

total surface area of a cylinder =	$2\pi r^2$	+	$2\pi rh$
	(area of the two circles)		(area of the rectangle)

Example 2

Q Calculate the total surface area of this solid cylinder. Leave your answer in terms of π.

[diagram: cylinder, height 12 cm, base 6 cm]

A Surface area $= 2\pi r^2 + 2\pi rh$

$= 2 \times \pi \times 3^2 + 2 \times \pi \times 3 \times 12$

$= 18\pi + 72\pi$

$= 90\pi$ cm^2

Example 3

Q If the volume of this cylinder is 205 cm^3, work out its height. Use $\pi = 3.142$ or the π button on your calculator.

Give your answer correct to three significant figures.

5.6 cm

A $V = \pi \times r^2 \times h$

$205 = \pi \times 2.8^2 \times h$

$h = \dfrac{205}{\pi \times 2.8^2}$

$= 8.3231...$

$h = 8.32$ cm

Example 4	**Q** Calculate: **a** the volume **b** the total surface area of the cone.

Give your answers correct to two decimal places.

A   **a** $V = \dfrac{1}{3}\pi r^2 h$

$\quad = \dfrac{1}{3} \times \pi \times 5^2 \times 12$

$\quad = 100\pi$

$\quad = 314.16 \text{ cm}^3$ (2 d.p.)

b $A = \pi r l + \pi r^2$

Slant height, $l = \sqrt{12^2 + 5^2}$

$\quad = \sqrt{144 + 25}$

$\quad = \sqrt{169}$

$\quad = 13 \text{ cm}$

$A = \pi r l + \pi r^2$

$\quad = \pi \times 5 \times 13 + \pi \times 5^2$

$\quad = 65\pi + 25\pi$

$\quad = 90\pi$

$\quad = 282.7433...$

$\quad = 282.74 \text{ cm}^2$ (2 d.p.)

> See Chapter 18 on Pythagoras' theorem.

Example 5	**Q** The mass of this solid toy is 637 g.

Work out the density of the toy. Give your answer correct to two decimal places.

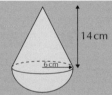

A First, work out the volume of the toy.

Volume of cone $= \dfrac{1}{3}\pi r^2 h$

$\qquad\qquad = \dfrac{1}{3} \times \pi \times 6^2 \times 14$

$\qquad\qquad = 527.787... \text{ cm}^3$

Volume of sphere $= \dfrac{4}{3}\pi r^3$

So volume of hemisphere $= \dfrac{2}{3}\pi r^3$

$\qquad\qquad\qquad\qquad = \dfrac{2}{3} \times \pi \times 6^3$

$\qquad\qquad\qquad\qquad = 452.389... \text{ cm}^3$

Total volume: $527.787... + 452.389... = 980.176... \text{ cm}^3$

Density $= \dfrac{\text{mass}}{\text{volume}}$

> See Chapter 14 for more density calculations.

$\qquad = \dfrac{637}{980.176...}$

$\qquad = 0.6498...$

$\qquad = 0.65 \text{ g/cm}^3$ (2 d.p.)

Practice questions

1 Calculate the volume of this sphere.

Use the formula: volume $= \frac{4}{3}\pi r^3$

All measurements are in centimetres.

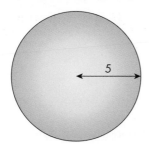

2 Calculate the volume of this square-based pyramid.

Use the formula: volume $= \frac{1}{3} \times$ base area $\times h$

All measurements are in metres.

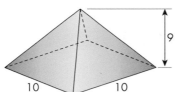

3 The diagram shows a solid cone.
Work out the total surface area.

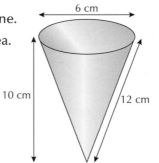

> **Hint**
>
> Curved surface area of cone $= \pi r l$, where l is the slant height.

4 The diagram shows a solid steel cylinder.
The density of steel is 7.85 g/cm³.
Work out the mass of the cylinder.
Give your answer correct to the nearest gram.

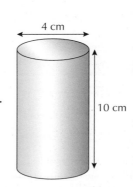

> **Hint**
>
> You need to learn this formula:
>
> Volume of a cylinder $= \pi r^2 h$
>
> where r is the radius and h is the perpendicular height.

5 The diagram shows an ice cream tub in the shape of a frustrum.

(A frustrum is part of a cone. Think of it as a large cone with a smaller cone removed.)

a Calculate the volume of ice cream it can contain when filled completely.

b Calculate the surface area of the ice cream tub. (Remember to include the area of the base.)

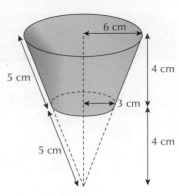

21.3 Converting units of area and volume

Converting units of area

Before you calculate an area, always check that the measurements are in the same units.

This square has side length 1 metre.

$1 \text{ m} = 100 \text{ cm}$

$1 \text{ m}^2 = 100 \text{ cm} \times 100 \text{ cm}$

$1 \text{ m}^2 = 10\ 000 \text{ cm}^2$

Example 6	**Q** Convert 8 m² into square centimetres.
	A $1 \text{ m}^2 = 10\ 000 \text{ cm}^2$ So $8 \text{ m}^2 = 8 \times 10\ 000 \text{ cm}^2$ $= 80\ 000 \text{ cm}^2$

Converting units of volume

This cube has a length of 1 m.

$1 \text{ m} = 100 \text{ cm}$

$1 \text{ m}^3 = 100 \text{ cm} \times 100 \text{ cm} \times 100 \text{ cm}$

$1 \text{ m}^3 = 1\ 000\ 000 \text{ cm}^3$

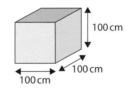
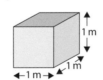

Example 7	**Q** Convert 12 m³ into cubic centimetres.
	A $1 \text{ m}^3 = 1\ 000\ 000 \text{ cm}^3$ So $12 \text{ m}^3 = 12 \times 1\ 000\ 000 \text{ cm}^3$ $= 12\ 000\ 000 \text{ cm}^3$

Exam tips Drawing a quick sketch will help you work out area and volume conversions.

Practice questions

1 Mary has a piece of card that measures 0.5 m by 20 cm.

She says the area of the card is 10 cm².

Is she correct?

You must explain your answer.

2 The top surface of a desk is a rectangle that measures 120 cm by 60 cm.

a Work out the area of the top surface of the desk in square centimetres.

b Convert your answer to part **a** into square metres.

3 One litre of water fits into a cube that measures 10 cm by 10 cm by 10 cm.

a Work out the volume of one litre of water in cubic centimetres.

b Convert your answer to part **a** to cubic metres.

4 Jane has 15 new windows fitted.

Each window has two pieces of glass that are 0.5 cm thick.

Each piece of glass is a rectangle measuring 50 cm by 150 cm.

a Calculate the volume of glass in Jane's new windows. Give your answer in cubic centimetres.

b Convert your answer to part **a** into cubic metres.

Glass has a density of 2500 kg/m³.

c Calculate the total mass of glass in Jane's new windows.

REVISION CHECKLIST

- Volume is measured in cubic units.
  - Volume of a cuboid, $V = lwh$
  - Volume of a prism, V = area of cros-section (A) × length (l)
  - Volume of a cylinder, $V = \pi r^2 \times h$
  - Volume of a sphere, $V = \frac{4}{3} \pi r^3$
  - Volume of a pyramid, $V = \frac{1}{3}$ × area of base × height (h)
  - Volume of a cone, $V = \frac{1}{3} \pi r^2 h$
- Surface area is measured in square units.
  - Surface area of a sphere = $4\pi r^2$
  - Curved surface area of a cone = $\pi r l$
  - Surface area of a cylinder = $2\pi r^2 + 2\pi r h$
- 1 m² = 10 000 cm²
- 1 m³ = 1 000 000 cm³

Exam tips The relevant formulae for the surface area and the volume of a cone and a sphere will be given with the exam question.

Exam-style questions

1 Calculate the volume of this cuboid.

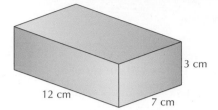

3 cm

12 cm

7 cm

2 Calculate the total surface area of this prism.

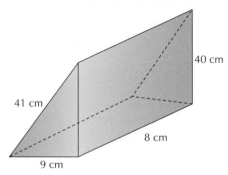

40 cm

41 cm

8 cm

9 cm

3 The diagram shows a channel in a block that is to be filled with concrete.

Calculate the volume of concrete required to fill the channel.

Give your answer in litres. (1 litre = 1000 cm^3)

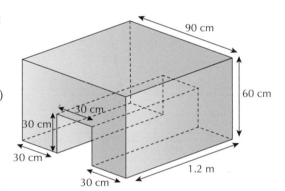

90 cm

60 cm

30 cm

30 cm

30 cm

30 cm

30 cm

1.2 m

4 The volume of this cuboid is 12 096 cm^3.

Calculate the height, h, of the cuboid.

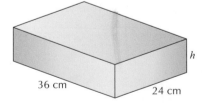

h

36 cm

24 cm

5 The total surface area of this cuboid is 236 cm^2.

Calculate the height, h, of the cuboid.

h

6 cm

8 cm

6 The diagram shows a cylindrical can of beans.

It contains 410 g of baked beans.

Work out the density of the contents of the can.

Give your answer correct to two decimal places.

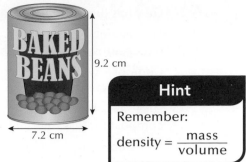

9.2 cm

7.2 cm

> **Hint**
>
> Remember:
>
> $\text{density} = \dfrac{\text{mass}}{\text{volume}}$

7 The diagram shows two views of a candle holder for two candles.

The candle holder is made from a solid block of wood in the shape of a cuboid with two identical cylindrical holes. The holes are 6 cm deep with diameter 8 cm.

Calculate the volume of the wood in the holder.

Give your answer correct to the nearest whole number.

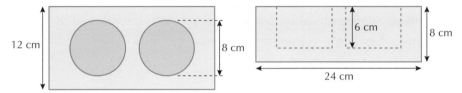

12 cm

8 cm

6 cm

8 cm

24 cm

8 Calculate the volume of the cuboid shown.

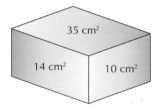

35 cm²

14 cm²   10 cm²

9 The space in the back of a van is in the shape of a cuboid, as shown.

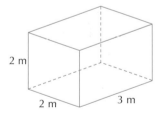

2 m

2 m   3 m

Boxes 40 cm by 50 cm by 30 cm are to be transported in the back of the van, as shown.

THIS WAY UP   30 cm

40 cm   50 cm

a What is the largest number of boxes that can be transported at one time?

b How would your answer to part **a** change if you could ignore the 'This way up'?

10 Lucy has just finished making a raised flower bed in her garden.

The diagram shows the shape and dimensions of the flower bed.

Compost is sold in bags containing 120 litres costing £7 each.

Lucy says, 'I can fill the flower bed with compost for less than £40.'

Is she correct?

Show your working.

(1 litre = 1000 cm³)

11 A shape has an area of 8730 cm². Convert this area into square metres.

12 A cuboid has a volume of 0.5 m³. Convert this volume to cubic centimetres.

13 Phil has an empty cylindrical oil tank. The dimensions are marked on the diagram.

The tank must not be filled to more than 85% of its total capacity.

Oil costs 45p per litre.

Phil says, 'It will cost more than £800 to fill the tank.'

Is he correct?

You must show all your working.

(1 litre = 1000 cm³)

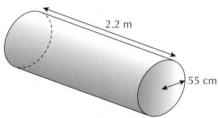

14 A milk jug holds 2 litres of milk when full.

Ella's glass is in the shape of a cylinder, with diameter 6 cm and height 8 cm.

How many times can she fill the glass from the jug?

(1 litre = 1000 cm³)

15 The model of a tower is built from modelling clay by joining a cuboid and a square-based pyramid, as shown.

The density of the modelling clay is 2.3 g/cm³.

Calculate the mass of the model.

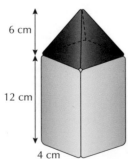

16 A capsule is in the shape of a cylinder with a hemisphere at each end.

Calculate the capacity of the capsule.

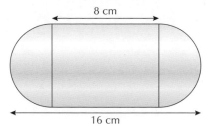

17 Jill has a cone made from modelling clay.

The cone has a base diameter of 16 cm and a slant height of 10 cm.

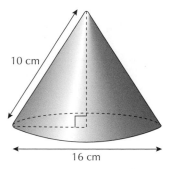

Jill reshapes this modelling clay into a cube.

Work out the length of one side of the cube correct to two decimal places.

> **Hint**
>
> Start by finding the height of the cone using Pythagoras' theorem.

18 Calculate the radius of a sphere that has volume 2000 cm^3.

Give your answer correct to two decimal places.

19 A cube with sides of length x cm and a sphere with radius 12 cm have the same volume.

Work out the value of x.

Give your answer correct to one decimal place.

20 A cube with sides of length 6 cm and a sphere with radius x cm have the same volume.

Work out the value of x.

Give your answer correct to two decimal places.

Now go back to the list of objectives at the start of this chapter.

How confident do you now feel about each of them?

22 Vectors and transformations

Objectives

Before you start this chapter, mark how confident you feel about each of the statements below:

	▶	▶▶	▶▶▶
I can translate a shape by a vector and use column vectors to describe a translation.			
I can transform 2D shapes using single reflections and describe the reflection on a coordinate grid.			
I can draw the position of a shape after a rotation.			
I can find the centre, angle and direction of rotation.			
I can enlarge a given shape from a point.			
I can describe enlarged 2D shapes using positive integer and fractional scale factors.			
I can find the centre of enlargement by drawing.			
I can compare areas and volumes, making links to similarity and scale factors.			
Stretch objectives Go to www.collins.co.uk/edexcelpost16			
I can calculate using column vectors and represent the sum and difference of two vectors graphically.			
I can identify column vectors that are parallel.			

Check-in questions

- Complete these questions to assess how much you remember about each topic. Then mark your work using the answers at the back of the book.
- If you score well on all sections, you can go straight to the Revision checklist and Exam-style questions at the end of the chapter. If you don't score well, go to the chapter section indicated and work through the examples and practice questions there.

1 Describe fully the transformation that maps:

   a A onto B   **b** B onto C   **c** A onto D   **d** A onto E.

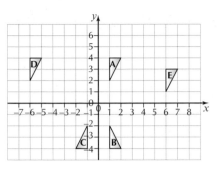

Go to 22.1, 22.2

2 The grid shows the positions of two triangles, A and B.

   a Reflect triangle A in the line $x = 5$.

     Label this image C.

   b Translate triangle B by the vector $\begin{pmatrix} 4 \\ 2 \end{pmatrix}$.

     Label this image D.

   c Describe fully the single transformation that maps triangle A onto triangle B.

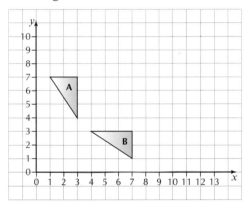

Go to 22.1, 22.2

3 Rotate triangle ABC 90° anticlockwise about the point (0, 1). Label the image A'B'C'.

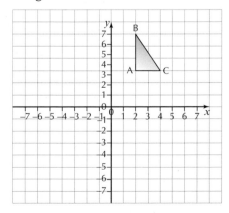

Go to 22.3

4 Copy the diagram and enlarge the shape shown by a scale factor of $\frac{1}{2}$ using centre of enlargement (0, 0). Label the shape T.

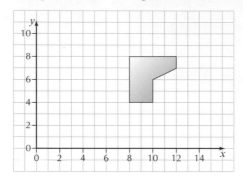

Go to 22.4

22.1 Translations

A **translation** is a **transformation** that moves a shape from one position to another. When a shape is translated, every point on the shape must be moved in the same direction and for the same distance. The object and the image are **congruent** (identical).

The horizontal and vertical displacement of a translation is described using a column vector, e.g. $\begin{pmatrix} a \\ b \end{pmatrix}$.

a represents the horizontal displacement and b represents the vertical displacement.

The column vector $\begin{pmatrix} 5 \\ 2 \end{pmatrix}$ means 5 units right and 2 units up.

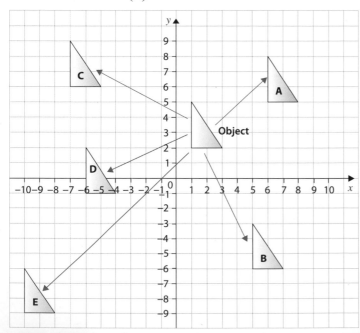

The object is translated by column vector:

$\begin{pmatrix} -8 \\ 4 \end{pmatrix}$ to give C

$\begin{pmatrix} -7 \\ -3 \end{pmatrix}$ to give D

$\begin{pmatrix} -11 \\ -11 \end{pmatrix}$ to give E

$\begin{pmatrix} 5 \\ 3 \end{pmatrix}$ to give A

$\begin{pmatrix} 4 \\ -8 \end{pmatrix}$ to give B.

Exam tips When describing this type of transformation, you must state it is a translation and write the associated column vector. Make sure you give the correct direction.
+ means right or up. – means left or down.

Practice questions

1   **a** Translate the shape by the vector $\begin{pmatrix} 3 \\ -1 \end{pmatrix}$.

     b Translate the shape by the vector $\begin{pmatrix} -1 \\ 2 \end{pmatrix}$.

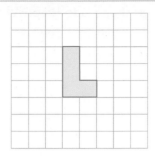

2   Describe the translation that maps:

     a A onto C

     b B onto C

     c B onto A

     d C onto A.

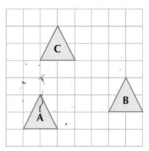

3   Amy says, 'The column vector for translating shape A to shape B is $\begin{pmatrix} 2 \\ 1 \end{pmatrix}$.'

Amy is wrong. What is the correct column vector for the translation?

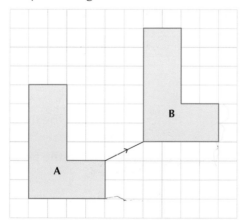

4   Write down the column vector for each translation.

     a E to C   **b** D to E

     c D to C   **d** C to D

     e A to C   **f** B to C

     g B to A   **h** B to E

     i B to D   **j** E to A

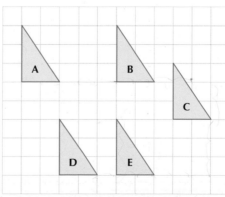

5 Write down the column vector for each translation.

    a shape D to shape A

    b shape E to shape D

    c shape C to shape B

    d shape A to shape D

    e shape B to shape A

    f shape A to shape C

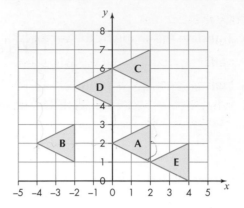

Exam tips   A common mistake in exam questions is to give the column vector that describes the translation the wrong way round. For example, from B to A instead of from A to B. Make sure you read the question carefully and check your answer.

22.2 Reflections

A **reflection** is a **transformation** that creates an image of an object in a mirror line.

The object and its image are **congruent**.

The mirror line is known as an **axis of reflection**.

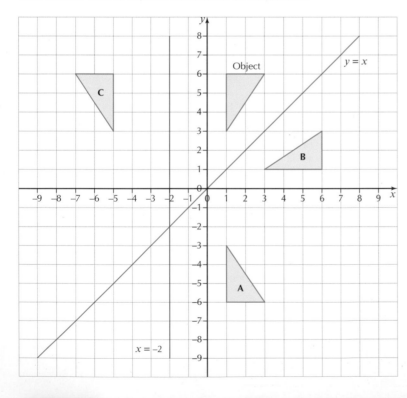

The object is:

- reflected in the x-axis ($y = 0$) to give image A.

- reflected in the line $y = x$ to give image B.

- reflected in the line $x = -2$ to give image C.

Exam tips   When describing this type of transformation, you must state it is a reflection and give the equation of the mirror line in the form '$y =$' or '$x =$'.

Practice questions

1 Copy the diagram.

  a Reflect triangle A in the *x*-axis. Label it B.

  b Reflect triangle A in the *y*-axis. Label it C.

  c Reflect triangle C in the *x*-axis. Label it D.

  d What do you notice about the triangles labelled B and D?

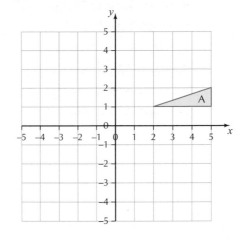

2 Reflect each shape in the line $x = 1$.

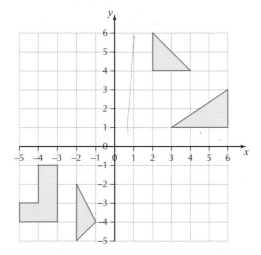

3 Reflect each shape in the line $y = 1$.

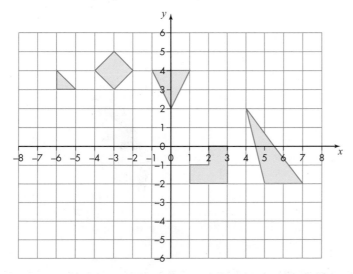

4   **a** Draw the line $y = x$ as a dashed line.

   b Draw the reflection of shape A in the line $y = x$. Label the shape W.

   c Draw the reflection of shape B in the line $y = x$. Label the shape X.

   d Draw the reflection of shape C in the line $y = x$. Label the shape Y.

   e Draw the reflection of shape D in the line $y = x$. Label the shape Z.

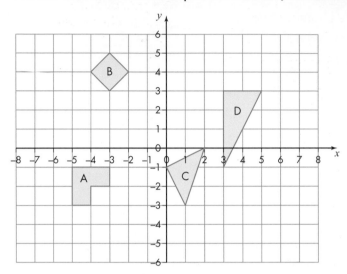

5   Describe the transformation that maps:

   a shape A to shape B

   b shape E to shape F

   c shape P to shape Q

   d shape X to shape Y

   e shape Y to shape X.

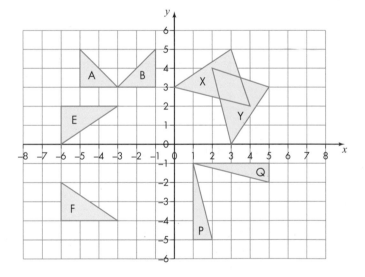

22.3 Rotations

A **rotation** is a **transformation** in which an object is rotated about a fixed point by an angle in either a clockwise or anticlockwise direction. The fixed point is called the **centre of rotation**. The size and shape of the figure are not changed, i.e. the image and the object are **congruent**.

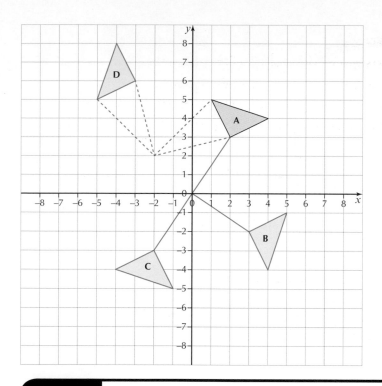

Object A is rotated:

- 90° clockwise about (0, 0) to give image B.

- 180° about (0, 0) to give image C.

- 90° anticlockwise (or 270° clockwise) about (–2, 2) to give image D.

> A rotation of 180° is the same clockwise and anticlockwise.

Practice questions

1   a   Identify the image after each rotation.

   i   shape A, 90° clockwise about O

   ii   shape D, 90° anticlockwise about O

   iii   shape B, 180° about O

   iv   shape B, 180° anticlockwise about O

   b   What is missing from this instruction:

   'To get from shape C to shape D, you rotate 90° about O'?

   c   Describe fully the transformation that maps shape C to shape B.

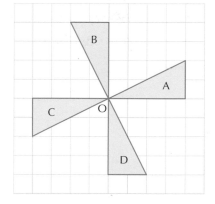

2   a   Draw the image of the shape after a rotation of quarter-turn clockwise about centre C.

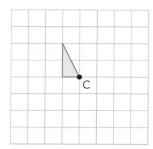

   b   Draw the image of the shape after a rotation of 180° about centre C.

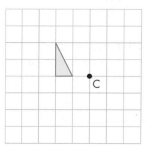

3 Draw the image of the shape after a rotation:

a 90° clockwise about the point (0, 0)

b 90° anticlockwise about the point (1, −3)

c 180° about the point (−1, −1).

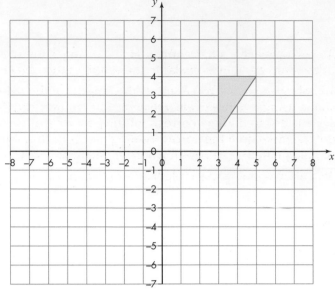

4 Rotate the shapes about centre (0, 0) to make a pattern with rotational symmetry of order 4.

a

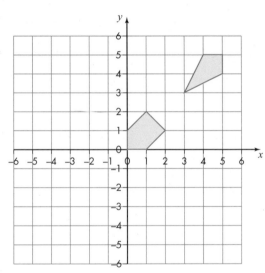

b

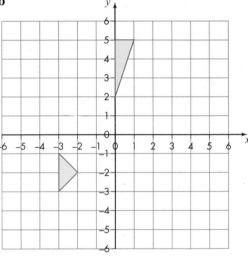

5 Describe fully the single transformation that maps triangle A on to triangle B.

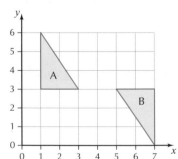

22.4 Enlargements

An **enlargement** is a **transformation** that changes the size but not the shape of the object, i.e. the shape of the enlargement is **similar** to the shape of the object.

To enlarge a shape, you multiply the lengths of the sides by a scale factor. When you enlarge a shape using a **centre of enlargement**, you multiply the distance from the centre to each vertex by the scale factor to find the corresponding positions of each vertex on the image. To find the centre of enlargement, join the corresponding points of the object and image, extending the lines. The centre of enlargement is the point where all the lines meet.

An enlargement with a scale factor between 0 and 1 makes the shape smaller.

| Example 1 | **Q** Describe fully the transformation that maps ABC onto A'B'C'. |

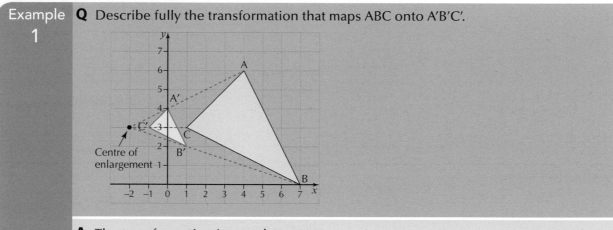

A The transformation is an enlargement.

When you join A to A', B to B'…, the lines meet at (–2, 3), so this is the centre of enlargement.

A'B' is one-third of the length of AB, so this is the scale factor.

The transformation is an enlargement by scale factor $\frac{1}{3}$ with centre of enlargement (–2, 3).

Exam tips When describing enlargements, remember to include the scale factor of the enlargement and the centre of enlargement.

Areas of enlarged shapes

If a shape is enlarged by scale factor k, then the area of the enlarged shape is k^2 times bigger.

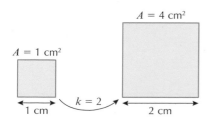

For example, if a square of side 1 cm is enlarged by a scale factor of 2, the sides of the enlarged shape are 2 times greater and the area of the enlarged shape is $2^2 = 4$ times greater.

See Chapters 13 and 15 for more on ratio and proportion and similarity.

Volumes of enlarged solids

If a shape is enlarged by scale factor k, then the volume of the enlarged shape is k^3 times bigger.

For example, if a cube of side 1 cm is enlarged by a scale factor of 2, the sides of the enlarged shape are 2 times greater and the volume of the enlarged shape is $2^3 = 8$ times greater.

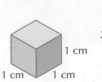

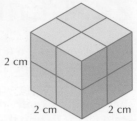

Practice questions

1 Enlarge the shape by a scale factor of 3 with (0, 0) as the centre of enlargement.

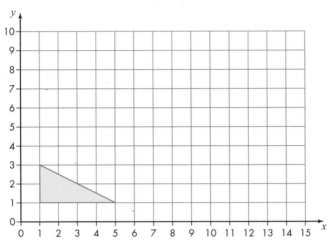

Hint

To get to point A on the shape, go across 1 and up 1. To get to the corresponding point on the image, go across 3 and up 3.

2 Enlarge each shape by a scale factor of 2, using C as the centre of enlargement.

a

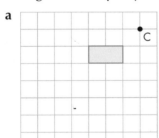

b

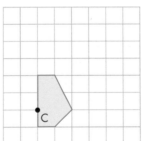

3 Enlarge each shape by the scale factor given, using C as the centre of enlargement.

a

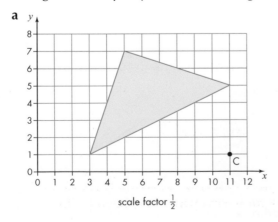

scale factor $\frac{1}{2}$

b

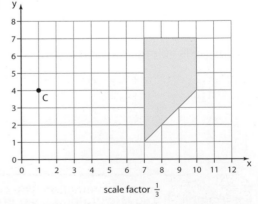

scale factor $\frac{1}{3}$

c

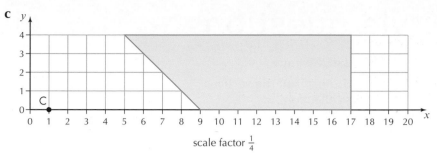

scale factor $\frac{1}{4}$

4 A triangle is enlarged by a given scale factor. For each part, work out the area of the enlarged shape.

  a Area of original triangle 5 cm², scale factor of enlargement 2

  b Area of original triangle 4 cm², scale factor of enlargement 3

  c Area of original triangle 24 cm², scale factor of enlargement 0.5

5 The diagram shows a small cube of side 2 cm and a larger cube of side 6 cm.

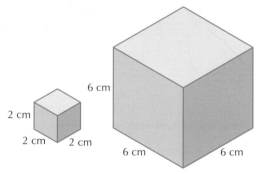

  a What is the scale factor of the enlargement?

  b Calculate the volume of the smaller cube.

  c What is the scale factor for the volume from the smaller cube to the larger one?

  d Use your answers to **b** and **c** to calculate the volume of the larger cube. Check your answer using the side length of the larger cube.

REVISION CHECKLIST

● A column vector is written as $\begin{pmatrix} a \\ b \end{pmatrix}$.

● Column vectors can be used to describe the displacement of translations.

● The object and image are congruent when a shape is translated or reflected. They are similar when a shape is enlarged.

● An enlargement with a scale factor between 0 and 1 makes the shape smaller.

Exam tips In the examination, you can ask for tracing paper to help you with transformation questions.

Exam-style questions

1 The kite is reflected in the line $x = -1$.

Write down the coordinates of the vertices of the kite after this reflection.

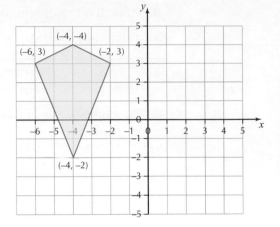

2 Reflect the shape in the line $y = 4$.

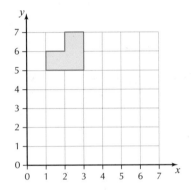

3 Reflect the trapezium in the line $y = x$.

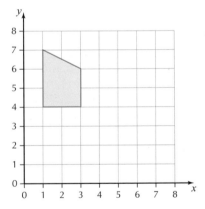

4 a Enlarge shape A by a scale factor of 2.

Label your new shape B.

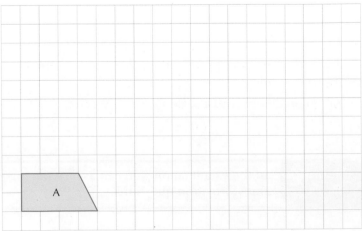

b Which of the following statements are true and which are false?

   i Shape B is similar to shape A.

   ii Shape B is congruent to shape A.

   iii The perimeter of shape B is twice the perimeter of shape A.

   iv The area of shape B is twice the area of shape A.

   v The angles in shape B are the same size as the angles in shape A.

   vi The angles in shape B are twice the size of the angles in shape A.

5 Enlarge this shape by a scale factor of 4.

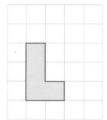

6 The parallelogram P has vertices at (–6, 4), (–2, 4), (–5, 6) and (–1, 6).

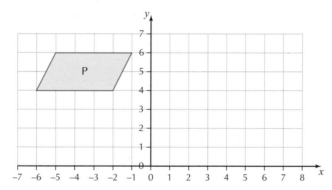

  a Draw the reflection of the parallelogram P in the line $y = 2$ and label it A.

  b Translate the parallelogram P 6 squares right and 2 squares down and label it B.

7 Enlarge the shape with scale factor 3, using point O as the centre of enlargement.

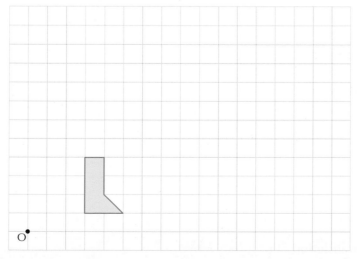

8   Enlarge the shape with scale factor $\frac{1}{2}$, using point O as the centre of enlargement.

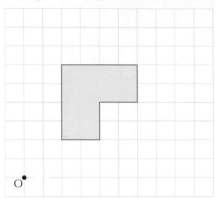

9   Shape A is rotated 90° anticlockwise about (0, 2) to shape B.

Shape B is rotated 90° anticlockwise about (0, 2) to shape C.

Shape C is rotated 90° anticlockwise about (0, 2) to shape D.

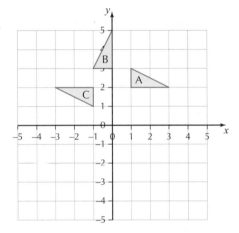

a Draw shape D on a copy of the diagram.

b Describe fully the single transformation that maps shape C to shape A.

10 Rhombus A can be transformed onto rhombus B using different transformations.

Copy and complete these sentences so that they describe fully one single transformation of rhombus A onto rhombus B.

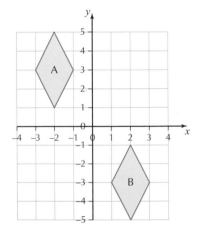

a Rhombus B is a translation of rhombus A by…

b Rhombus B is a rotation of rhombus A of…

11   **a** Rotate triangle P 180° clockwise about (0, 0).

Label your new triangle Q.

b Enlarge triangle A by a scale factor $\frac{1}{2}$, centre (0, 0).

Label your new triangle R.

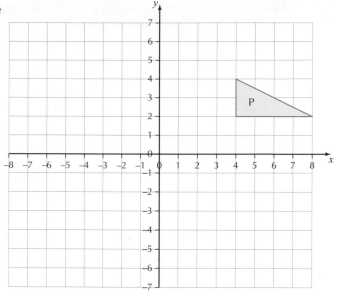

12   Triangle PQR is an enlargement of triangle ABC with scale factor $\frac{8}{5}$.

Work out the length PR.

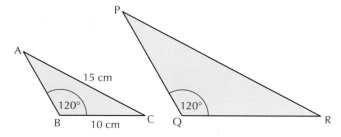

13   True or false?

   a All squares are enlargements of each other.

   b All rectangles are enlargements of each other.

   c All equilateral triangles are enlargements of each other.

   d All parallelograms are enlargements of each other.

   e All isosceles triangles are enlargements of each other.

   f All regular hexagons are enlargements of each other.

14

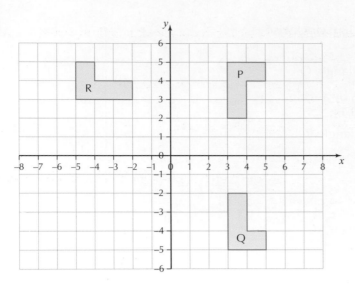

a Describe fully the single transformation that takes shape P to shape Q.

b Describe fully the single transformation that takes shape P to shape R.

c Translate P by the vector $\begin{pmatrix} 2 \\ -5 \end{pmatrix}$. Label the new shape T.

15 The diagram shows two 'L' shapes, A and B.

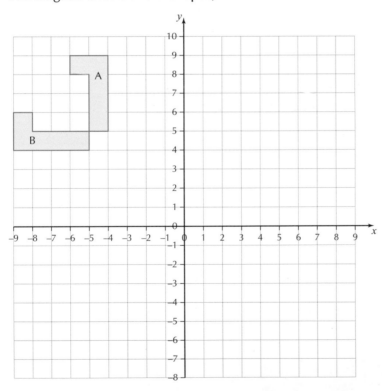

a Describe fully the single transformation that takes shape A to shape B.

b Shape A is translated by the vector $\begin{pmatrix} 3 \\ -7 \end{pmatrix}$ and then rotated 90° clockwise about the point (–1, –2) to give shape C. Describe fully the single transformation that maps shape A onto shape C.

16 Sarah prints out a borderless 10 cm by 15 cm photograph on her printer.

    a Work out the area of her photo.

    b She decides to print an enlargement of her photograph.
       She uses a scale factor of 1.5.
       Calculate the area of the enlarged photograph.

17 Mel has two bottles which are mathematically similar in shape.

The larger bottle has a volume of 2400 cm^3. It is twice the height of the smaller bottle.

Calculate the volume of the smaller bottle.

Now go back to the list of objectives at the start of this chapter.

How confident do you now feel about each of them?

23 Probability

Objectives

Before you start this chapter, mark how confident you feel about each of the statements below:

	▶	▶▶	▶▶▶
I am able to use words associated with probability, e.g. *fair*, *even*, *certain*, *unlikely*, etc.			
I can represent and compare probabilities on a probability scale from 0 to 1.			
I can use probabilities written as fractions, decimals and percentages.			
I list the outcomes of single events and find their probabilities.			
I understand the terms *mutually exclusive* and *exhaustive*.			
I can find the missing probability from a table.			
I know that the P(event not happening) = 1 – P(event happening).			
I can use relative frequency to estimate probability.			
I can estimate the number of successes based on probability.			
I understand that a larger number of trials results in a better estimate of probability.			
I can represent outcomes of two events (e.g. throwing two coins) in a sample space diagram or a list.			
I can apply systematic listing strategies including the product rule for counting.			
I understand what is meant by *independent* and *conditional* probability.			
I can use frequency trees to find the probabilities associated with single events.			
I can draw a tree diagram to show probabilities of two events.			

Check-in questions

- Complete these questions to assess how much you remember about each topic. Then mark your work using the answers at the back of the book.
- If you score well on all sections, you can go straight to the Revision checklist and Exam-style questions at the end of the chapter. If you don't score well, go to the chapter section indicated and work through the examples and practice questions there.

1 Copy and complete each statement using one of the probability words in the box.

> impossible   very unlikely   unlikely   evens
> likely   certain

a It is _____ that 15 people in the same class have the same birthday.

b It is _____ that the Prime Minister will have two heads tomorrow.

c It is _____ that most people will wear a coat when it is raining.

Go to 23.1

2 A drawer contains six red and five blue socks. If a sock is chosen at random from the drawer, work out the probability that the sock is:

a red   **b** blue   **c** red or blue   **d** yellow   Go to 23.2

3 The probability that Sarah sends a text on any given day is 0.64. What is the probability that Sarah does not send a text on any given day?

Go to 23.3

4 The probability that Nisha buys a chocolate bar is $\frac{7}{9}$. What is the probability that Nisha does not buy a chocolate bar?   Go to 23.3

5 A dice was thrown 320 times. A score of 4 was obtained 58 times. What is the relative frequency of the score 4?   Go to 23.4

6 Reece has a pizza for his lunch. He has a choice of three toppings – mushroom, pineapple or ham. He chooses two toppings. Make a list of all the combinations he can choose.   Go to 23.5

7 Two fair dice are thrown at the same time. The total score is the product of the two scores.

a Draw a sample space diagram to show the total scores.

b Work out the probability of:

i a score of 3   **ii** a score that is a multiple of 5.   Go to 23.5

8 The two-way table shows the numbers of infants who were immunised against an infectious disease and the numbers of infants who caught the disease.

	Immunised	Not immunised	Total
Did not catch the disease	83	8	91
Caught the disease	4	17	21
Total	87	25	112

a What is the probability that an infant who has been immunised catches the disease?

b What is the probability that an infant has been immunised?

Go to 23.5

9 Sangeeta has a biased dice. The probability it lands on 3 is 0.4. She rolls the dice twice.

a Copy and complete the tree diagram.

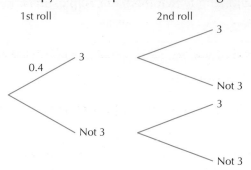

b Work out the probability that she throws:

i two 3s   **ii** exactly one 3.

Go to 23.6

23.1 The probability scale

Probability is the chance or likelihood that something will happen. All probabilities lie between 0 (impossible) and 1 (certain). They can be written as fractions, decimals or percentages.

They can be shown on a probability scale.

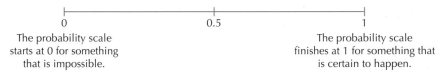

The probability scale starts at 0 for something that is impossible.

The probability scale finishes at 1 for something that is certain to happen.

Example 1	**Q** A bag contains three red beads, three blue beads and six green beads. Mark these probabilities on a probability scale.

a P, the probability of choosing a green bead

b T, the probability of choosing a red bead

c X, the probability of choosing a black bead

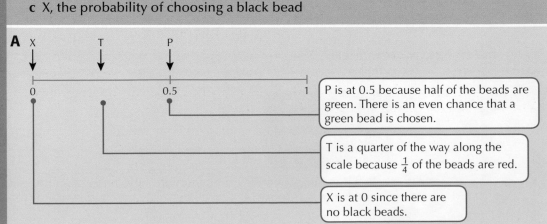

P is at 0.5 because half of the beads are green. There is an even chance that a green bead is chosen.

T is a quarter of the way along the scale because $\frac{1}{4}$ of the beads are red.

X is at 0 since there are no black beads.

Practice questions

1 A bag contains three red balls, five blue balls and two green balls. A ball is chosen at random. Work out the probability that the first ball picked is:

   a red   **b** blue   **c** yellow.

2 Hannah has worked out that the probability that she is late for work is 1.2.

Explain how you know she is wrong.

3 What is the probability that you will have had your birthday for this year by December 31st?

4 A bag of chocolates contains one strawberry cream, three mint chocolates and two fudge. Draw a probability scale from 0 to 1 and mark the probability of each outcome when a chocolate is picked at random from the bag.

   a picking a strawberry cream   **b** picking a fudge or mint chocolate

   c picking a nut

5 Mia buys some popcorn at the cinema. She is asked if she would like sweet or salted. She tells the cashier to surprise her. The cashier picks one of the bags shown in the picture at random. Is Mia more likely to get sweet or salted popcorn? Explain your answer.

23.2 Calculating probabilities associated with single events

- Probability an event happens $= \dfrac{\text{number of ways an event can happen}}{\text{total number of outcomes}}$

- **Exhaustive events** account for all possible outcomes. For example, 1, 2, 3, 4, 5, 6 are all possible outcomes when a dice is thrown. The total probability of exhaustive events is 1.

- **Mutually exclusive** events are events that cannot happen at the same time, e.g. throwing a head and a tail on a fair coin cannot appear at the same time.

Example 2

Q The letters in the word MATHEMATICS are placed in a container, and a letter is taken out at random. What is the probability that it is: **a** a T **b** an S **c** an R?

A **a** $P(T) = \frac{2}{11}$

> There are 11 letters, so each probability is out of 11.

b $P(S) = \frac{1}{11}$

c $P(R) = 0$

> There is no letter R, so the probability is zero.

Exam tips Remember that probabilities must be between 0 and 1 inclusive, so if your answer is negative or greater than 1, check again!

Practice questions

1. Olivia chooses one letter from her name at random. What is the probability that she chooses: **a** v **b** a vowel **c** U?

2. Eddie delivers newspapers. He deliver six *Daily Mails*, three *Mirrors*, two *Guardians* and one *Financial Times*.

   Unfortunately, he forgets the delivery sheet that tells him which paper to deliver to each address so he delivers a newspaper to each house at random.

   a Bert ordered *The Daily Mail*. What is the probability that he received the correct newspaper?

   b Daz ordered *The Mirror*. What is the probability that he received the correct newspaper?

3. A card is taken at random from a shuffled pack of ordinary playing cards.

   a Find the probability that the card is:

   i red **ii** a heart **iii** the 10 of clubs.

   b Explain why your answers to parts **ii** and **iii** are mutually exclusive.

4. Give an example of two mutually exclusive events.

5. Louise is playing a word game. She gets these tiles:

   O₁ Q₁₀ Z₁₀ A₁ E₁ X₈ L₁

   Whilst waiting for her turn, she plays about with her tiles, pulling them out of the box at random. What is the probability that she pulls out:

   a Z **b** a consonant **c** J **d** a consonant or a vowel?

23.3 Probability that an event will not happen

The probabilities of an event happening and not happening are mutually exclusive and exhaustive so the probabilities add up to 1. This means that when you know the probability that an event will happen, you can calculate the probability that the event will not happen using:

P(event will not happen) = 1 – P(event will happen).

Example 3	**Q** The probability that an alarm clock does not go off is 0.21. What is the probability that the alarm clock does go off?

A P(goes off) = 1 – P(does not go off)

$$= 1 - 0.21$$
$$= 0.79$$

Example 4	**Q** The probability that a 17-year-old passes their driving test on the first attempt is $\frac{3}{7}$. What is the probability they do not pass their driving test on the first attempt?

A P(fail) = 1 – P(pass)

$$= 1 - \frac{3}{7}$$
$$= \frac{4}{7}$$

Example 5	**Q** Imran throws a dart at a target. The table shows information about the probability of each possible score.

Score	0	1	2	3	4	5
Probability	0.04	x	$4x$	0.23	0.19	0.28

Imran is four times more likely to score 2 points than to score 1 point.

Work out the value of x.

A $0.04 + x + 4x + 0.23 + 0.19 + 0.28 = 1$

$$5x + 0.74 = 1$$
$$5x = 1 - 0.74$$
$$= 0.26$$
$$x = 0.052$$

Practice questions

1 Ellie wants to go to a night club. The probability that she gets in is 0.2.

What is the probability that she does not get in?

2 Daisy is a children's nurse. The probability that her next patient needs a vaccination is 0.05.

What is the probability that her next patient does not need a vaccination?

3 Copy and complete the probability table for this biased dice.

Score	1	2	3	4	5	6
Probability	0.3	0.1	0.04		0.45	0.1

4 The outcomes are exhaustive. Work out the value of x.

Colour	Red	Yellow	Blue	Green	Black
Probability	$2x$	0.13	$3x$	0.31	0.06

5 Josh is a forklift-truck driver. He can be sent to one of two warehouses. The probability that he is sent to Warehouse A is $\frac{4}{9}$. What is the probability that he is sent to Warehouse B?

23.4 Experimental probability

Experimental probability is determined by the results of a number of trials or events. It is called **relative frequency**.

$$\text{Relative frequency of an event} = \frac{\text{number of favourable outcomes}}{\text{total number of trials}}$$

For example, a dice is rolled 55 times. The score is 4 in 13 of these trials. The relative frequency of the score 4 is $\frac{13}{55}$.

The more times the experiment is carried out, the closer the relative frequency is to the theoretical probability. (Theoretical probability analyses a situation mathematically.)

Example 6	**Q**	The frequency table shows the results of throwing a fair coin 100 times. What is the estimated probability of a head?

Result	Tally	Frequency
Head	ⵘⵘ ⵘⵘ ⵘⵘ ⵘⵘ ⵘⵘ ⵘⵘ ⵘⵘ ⵘⵘ ⵘⵘ |||	48
Tail	ⵘⵘ ⵘⵘ ⵘⵘ ⵘⵘ ⵘⵘ ⵘⵘ ⵘⵘ ⵘⵘ ⵘⵘ ⵘⵘ ||	52

A Estimated probability $= \dfrac{\text{number of favourable outcomes}}{\text{total number of trials}}$

$$= \frac{48}{100}$$

$$= 0.48$$

The estimated probability of a head is 0.48.

You can use probability to predict the outcome of an event.

Expected number of outcomes = number of trials × probability

Example 7	**Q** The probability of passing a biology exam is 0.73.
	If 100 people take the exam, how many would you expect to pass?
	A 100 × 0.73 = 73 people

Example 8	**Q** The probability that Alex is late for work is 0.2.
	He has 40 working days until his next holiday.
	On how many days can Alex's boss expect him to be late before his next holiday?
	A P(late) is 0.2 × 40 = 8 days
	Alex's boss can expect him to be late for work on 8 days.

Exam tips Always check that your answers seem reasonable and make sense.

Practice questions

1 A fair coin is thrown 1000 times. How many times would you expect the outcome to be a head.

2 I have a bag containing 30 balls; 15 blue, 10 red and 5 green. I take a ball at random, note its colour and put it back in the bag. I do this 300 times. How many times would I expect to get:

a a red ball

b a blue or a green ball

c a ball that is not green

d a yellow ball?

3 Crawford and Janet each carry out an experiment using the same dice. The tables show their results.

Crawford's results

Number	1	2	3	4	5	6
Frequency	6	4	13	9	10	8

Janet's results

Number	1	2	3	4	5	6
Frequency	53	48	45	54	50	52

Crawford thinks the dice is biased. Janet thinks the dice is fair. Who is correct? Explain your choice.

4 Niamh rolls a dice and records the number of times she scores 3 throughout the experiment.

Number of throws	10	50	100	200	500	1000
Number of 3s	3	6	19	30	92	163
Experimental probability						

  a Copy and complete the table.

  b What is the theoretical probability of the score 3?

  c If Niamh threw the dice 6000 times, how many times should she expect to score 3?

5 Liam gets the train to his work placement at the fire station. The probability that the train is late on any one day is 0.35.

He needs to be at the fire station on the dates circled on the calendar

On how many of these days can Liam expect to be late?

JULY						
S	M	T	W	T	F	S
						1
2	3	4	5	6	7	8
9	10	11	12	13	14	15
16	17	18	19	20	21	22
23	24	25	26	27	28	29
30	31					

23.5 Possible outcomes of two or more events

You need to be able to find all the possible outcomes of multiple events in order to calculate the probability of any one of these outcomes. You can do this using an **ordered list**, a **sample space** diagram or a **two-way table**. You can also use the product rule for counting to find the total number of possibilities without writing them all down.

Lists

You must write lists in an ordered way to make sure that you include all the possibilities.

The meal combinations for a two-course meal from this menu are:

pizza, yogurt   chicken pie, yogurt   fish, yogurt

pizza, cake   chicken pie, cake   fish, cake

$P(\text{pizza, cake}) = \frac{1}{6}$

MENU

Pizza
Chicken pie
Fish
~~
Yogurt
Cake

Product rule for counting

The product rule for counting can also be used to find the total number of possible combinations. For example, if there are *A* ways of doing task 1 and *B* ways of doing task 2, there are *A* × *B* ways of doing both tasks.

Example 9	**Q** How many different meal combinations (main course and dessert) can be chosen from the menu above?
	A There are three different main courses (pizza, chicken pie and fish) and two different desserts (yogurt and cake). So there are 3 × 2 = 6 different combinations.

Two-way tables

Example 10

Q The two-way table shows the genders and handedness of students in a class.

Handedness	Male	Female	Total
Right	14	10	24
Left	2	7	9
Total	16	17	33

a What is the probability that a person chosen at random is right-handed?

b If a boy is chosen at random, what is the probability that he is left-handed?

A a P(right-handed) is $\frac{24}{33}$

b P(boy is left-handed) is $\frac{2}{16} = \frac{1}{8}$

Sample space diagrams

Sample space diagrams are particularly useful when there are several options for each event.

Example 11

Q Two spinners are spun and the scores added.

a Show the outcomes on a sample space diagram.

b What is the probability of a total score of 4?

c What is the probability that the total score is an odd number?

A a

		Spinner 1		
		1	2	3
Spinner 2	2	3	4	5
	3	4	5	6
	3	4	5	6

b P(4) is $\frac{3}{9} = \frac{1}{3}$

c P(odd) is $\frac{4}{9}$

1
  a List the possible outcomes when three coins are thrown at the same time, for example, HHH.

  b What is the probability of throwing three heads?

  c What is the probability of throwing two heads and one tail?

2
Two four-sided dice in the shape of a triangular-based pyramid are thrown at the same time and their scores are added.

  a Draw a sample space diagram showing all the possible outcomes.

  b What is the probability that the total score is 6?

3
A coin is tossed and a four-sided dice is rolled at the same time. What is the probability of getting a tail on the coin and a 2 on the dice?

4
The table shows the probabilities that students of a college are vegetarian by gender.

	Male	Female
Vegetarian	0.07	0.2
Not vegetarian	0.42	0.31

  a What is the probability that a student chosen at random is a vegetarian?

  b The college has 170 female vegetarians. How many students attend the college?

  c How many of the students are males that are not vegetarian?

5
A fair six-sided dice is rolled twice. What is the probability of rolling a 1 and a 2 in any order?

23.6 Probabilities of multiple events and tree diagrams

There are two important rules for calculating probabilities of multiple events depending on the type of events. Remember that two or more events are **independent** when the outcome of the second event is not affected by the outcome of the first event.

- **The OR rule**

  If two events are **mutually exclusive**, the probability of A or B happening is found by adding the probabilities.

  P(A or B) = P(A) + P(B)

- **The AND rule**

  If two events are **independent**, the probability of A and B happening together is found by multiplying the separate probabilities.

  P(A and B) = P(A) × P(B)

Frequency trees are used to show the outcomes of two events.

Example 12

Q A shop employs 80 people. 40% work part time, the rest work full time.
17 part-time staff and 27 full-time staff are male.

a Draw a frequency tree to show this information.

b What is the probability that a worker chosen at random is female?

A a

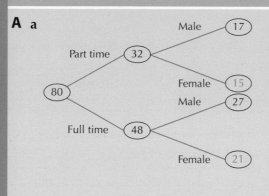

Fill in the information that is given.

Work out the number of part time and full time workers.
$0.4 \times 80 = 32$
$0.6 \times 80 = 48$

Work out the number of part time and full time female workers.
$32 - 17 = 15$
$48 - 27 = 21$

b P(female) is $\dfrac{36}{80} = \dfrac{9}{20}$

You can use **tree diagrams** to show the outcomes of events and to support calculating probabilities.

The probabilities on the branches leaving each point on the tree must add up to 1.

Example 13

Q A bag contains three red counters and four blue counters. A counter is taken from the bag at random. Its colour is noted and it is put back in the bag. A second counter is then taken out of the bag.

a Draw a probability tree diagram to represent these events.

b Work out the probability of picking: **i** two blues   **ii** one of each colour.

A a

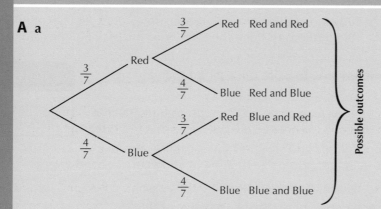

b **i** To find the probability of picking a blue AND a blue, multiply along the branches.

$$P(B) \times P(B) = \frac{4}{7} \times \frac{4}{7}$$

$$= \frac{16}{49}$$

(continued)

23.6 Probabilities of multiple events and tree diagrams

ii This is P(blue, red) OR P(red, blue), so you need to use the AND rule and the OR rule.

P(blue, red) is $P(B) \times P(R) = \frac{4}{7} \times \frac{3}{7}$

$$= \frac{12}{49}$$

P(red, blue) is $P(R) \times P(B) = \frac{3}{7} \times \frac{4}{7}$

$$= \frac{12}{49}$$

So P(one of each colour) is $\frac{12}{49} + \frac{12}{49} = \frac{24}{49}$

If one event depends on the outcome of another event, the two events are **dependent events**.

A **conditional probability** is the probability of a dependent event. The probability of the second outcome depends on what has already happened in the first outcome.

If the counter had not been replaced in the example above, the tree diagram would look like this:

This time the probability of picking two blues would be

$$P(B) \times P(B) = \frac{4}{7} \times \frac{3}{6} = \frac{12}{42}$$

Exam tips   Do not simplify the fractions on the branches of the tree. They will be easier to add with the same denominators.

Practice questions

1   **a**   A 10p coin and a 50p coin are thrown at the same time. Draw a probability tree to show all the possible outcomes. Remember to write the probabilities along each branch.

   b   Use your probability tree to work out each probability.

     i   P(HH)   **ii**   P(HT)

     iii   P(one head and one tail)   **iv**   P(at least one head)

> **Hint**
>
> P(at least one head) = 1 – P(no heads)

2   A bag contains four toffees and six chocolates. All the sweets look identical. Adrian takes and eats a sweet at random. He then takes and eats a second sweet.

   a   Draw a probability tree to show all possible outcomes.

   b   Use your probability tree to work out each probability.

     i   P(Adrian picks two toffees)   **ii**   P(Adrian picks at least one toffee)

     iii   P(Adrian does not pick a toffee)

3   50 students practise taking penalties.

It was predicted that 30 students would score a goal. Of those students, 24 actually scored a goal.

In total, 32 students scored a goal.

Copy and complete the frequency tree.

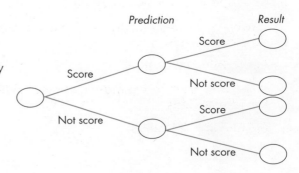

4 400 students took a test.

320 were predicted to pass the test.

10 of those predicted not to pass did actually pass.

310 students passed the test altogether.

a Copy and complete the frequency tree.

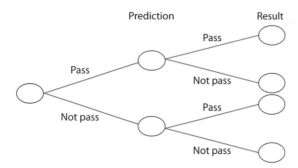

Prediction   Result

Pass

Pass

Not pass

Pass

Not pass

Pass

Not pass

b What percentage of students did not pass?

5 Hamish is playing in a football match and then in a cricket match.

The probability that his football team wins is $\frac{7}{10}$. The probability that his cricket team wins is $\frac{2}{5}$.

a Draw a probability tree to show all possible outcomes.

b Use your probability tree to work out the probability that:

i he is on the winning team for both matches

ii he is on the winning team for at least one match.

REVISION CHECKLIST

- All probabilities lie between 0 and 1.

- $P(\text{event will happen}) = \dfrac{\text{number of ways an event can happen}}{\text{total number of outcomes}}$

- P(event will not happen) = 1 − P(event will happen)

- Relative frequency = $\dfrac{\text{number of succesful outcomes}}{\text{total number of trials}}$

  Expected number of outcomes = number of trials × probability

- Mutually exclusive events cannot happen at the same time.

  If two events are mutually exclusive, the probability of A or B happening is found by adding the probabilities.

  P(A or B) = P(A) + P(B)

- If two events are independent, the probability of A and B happening is found by multiplying the probabilities.

  P(A and B) = P(A) × P(B)

Exam-style questions

1 A tin contains eight coffee, six praline, four honeycomb and two fondant chocolates. Copy and complete the probability scale to show the probability of each of these being chosen at random.

   a honeycomb   **b** coffee   **c** fondant or coffee   **d** strawberry

```
0                                    1
```

2 From a standard deck of cards, is it more likely that you will randomly choose a picture card or a number card? Give a reason for your answer.

3 A holiday company offers apartments, hotel rooms and all-inclusive packages. They surveyed 100 passengers on a flight about which package they had booked. The probability scale shows some of the responses.

```
0                                    1
```

   a What is P(A), an apartment

   b What is P(B), an all-inclusive package?

4 The probability that Shona gets an interview for a job is 0.32.

   What is the probability that she does not get an interview?

5 The probability that a seed germinates is 0.621.

   Write down the probability that the seed does not germinate.

6 Tara has these letter cards.

A L G O R I T H M

   What is the probability that a letter chosen at random is:

   a a vowel   **b** not a T   **c** a letter from the first half of the alphabet?

7 Abisha is playing a game with a dice. She needs to score 12 to reach the end. On her first throw, she scores 6. Abisha says she cannot win on her next throw because she's just thrown a 6. Is she correct? Explain your answer.

8 The probability that Caleb returns his library book on time is 0.18.

   What is the probability that he does not return his library book on time?

9 Aria works in a café which sells hot drinks. The probability that the next customer orders something other than a hot chocolate is 0.764.

   What is the probability that the next customer orders a hot chocolate?

10 The probability of an event happening is 0.21.

What is the probability of the event not happening?

11 Kirsty and Emma each throw a coin a number of times. The table shows the outcomes.

	Heads	Tails
Emma	152	148
Kirsty	96	104

Does either of them have a biased coin? Explain your answer.

12 There are some green socks and some red socks in a drawer. Phil randomly pulls out one sock at a time and then replaces it. He does this 150 times and gets 48 red socks. What is the relative frequency of getting a red sock?

13 Troy records the number of times the traffic lights outside college are red when he reaches them. On 20 trips, the traffic light is red 13 times.

During the year, he will drive to college 120 times. How many of these times should he expect the traffic light to be red?

14 Sarah runs a community café in her village. Her morning special is shown on the right.

The hot drinks available are tea, coffee and hot chocolate. The cakes available are chocolate, apple, toffee and parkin.

List all the possible choices for this offer.

> Any hot drink plus cake
> £3.50

15 Toby wants to have breakfast at the work canteen after his night shift. He can choose from apple juice or orange juice and from toast, cereal or fruit.

a How many different combinations are possible?

b He asks his friend to get his breakfast for him, but doesn't tell him what to get. What is the probability that he gets apple juice and cereal?

16 The table shows some information about the numbers of children in a nursery that have been vaccinated against MMR.

	Girl	Boy	Total
Vaccinated		39	87
Not vaccinated		13	
Total	54		106

a Copy and complete the table.

b What is the probability that a child chosen at random has been vaccinated against MMR?

17 The table shows the information about the people in a factory who have a driving licence.

	Male	Female
Has driving licence	96	126
Does not have driving licence	30	48

a What is the probability that a person chosen at random is a male without a driving licence? Give your answer as a fraction in its simplest form.

b A woman is selected at random. What is the probability that she has a driving licence?

18 150 people in the canteen were asked how they had travelled to college that day.

Of the 65 males, 13 had taken the bus and 38 had travelled by car.

27 females had travelled by car and 51 people had walked.

Copy and complete the frequency tree.

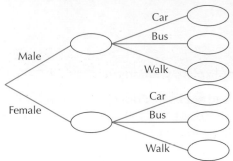

19 75 people were asked whether they lived in England or Scotland.

Of the 58 women, 44 were from Scotland.

26 people were from England.

Copy and complete the frequency tree.

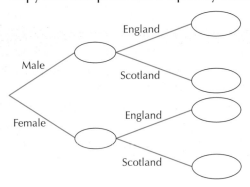

20 The information in the two-way table shows the drinks available in the vending machine at a leisure centre.

	Regular	Sugar free
Cola	24	1
Lemonade	20	9

Represent this information in a frequency tree.

**Now go back to the list of objectives at the start of this chapter.
How confident do you now feel about each of them?**

24 Sets and Venn diagrams

Check-in questions

- Complete these questions to assess how much you remember about each topic. Then mark your work using the answers at the back of the book.

- If you score well on all sections, you can go straight to the Revision checklist and Exam-style questions at the end of the chapter. If you don't score well, go to the chapter section indicated and work through the examples and practice questions there.

1. Set C = {2, 3, 4, 5, 6, 7} and Set D = {6, 7, 8, 9}

   Write down these sets.

   a $C \cup D$ b $C \cap D$

   Go to 24.1

2. Set A = {1, 2, 3, 5, 7, 9} and Set B = {1, 3, 5, 7, 9, 11}

   Write down these sets.

   Go to 24.1

   a $A \cup B$ b $A \cap B$ c $(A \cap B)'$

   Go to 24.1

3. The Venn diagram shows the subjects studied by 60 students.

   Of these, 6 study only Geography, 20 study both Geography and History and 16 do not study either subject.

   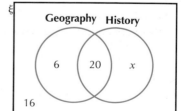

   a Copy and complete the Venn diagram.

   b Work out the value of x.

   c How many students study Geography?

   Go to 24.2

   d What is the probability that a student chosen at random does not study Geography or History?

   Go to 24.3

24.1 Set notation

- A **set** is a defined collection of objects. These are called the elements or members of the set.

  For example, Set A = {1, 2, 3 … 100}. This set is all the integers from 1 to 100.

- A **subset** is a set made from elements of a larger set.

  For example, Set B = {2, 4, 6 … 100}. This set is all the even numbers from 1 to 100 and is a subset of set A.

- A **finite set** is a set with a finite (limited) number of elements.

  For example, Set C = {2, 4, 6 … 20}. These are the even numbers up to 20.

- An **infinite set** is a set with an infinite (unlimited) number of elements.

  For example, Set D = {2, 4, 6 …}. These are all the even numbers.

- An **empty** or **null set** Ø is a set that contains no elements.

- A **universal set** ξ is a set that contains all possible elements.

- A′ is the set of elements *not* in set A.

Union of sets (∪)

Set A ∪ B contains elements belonging to A or B or both.

Example 1	**Q** Set A = {2, 3, 4} and Set B = {4, 5, 6, 7}
	Write the set A ∪ B.
	A A ∪ B = {2, 3, 4, 5, 6, 7}

Intersection of sets (∩)

Set A ∩ B contains elements belonging to both A and B.

If A and B have no elements in common, then A ∩ B = Ø, i.e. the empty set.

Example 2	**Q** Set A = {2, 3, 4} and Set B = {1, 3, 5}
	Write the set A ∩ B.
	A A ∩ B = {3} since this is the only number in both sets.

Practice questions

1. Match each symbol to its meaning.

   A′ Union of sets

   Ø   Not A

   ∪   The empty set

   ∩   The intersection of sets

2 Set A = {A, B, C, X, Y, Z}

Which of these are subsets of set A?

| Set B = {A, B, C}   Set C = {A, B, Z}   Set D = {A, B, C, D}   Set E = {D, E, F} |

3 Set A = {red, blue, green} and Set B = {red, blue, green, yellow}

Which statement is true?

a Set A is a subset of Set B.

b Set B is a subset of Set A.

c $A \cup B$ = {red, blue, green}

d Set A is an infinite set.

4 Set C = {1, 2, 3, 4, 5, 6, 7, 8, 9, 10} and Set D = {2, 4, 6, 8, 10 …}

Is Set C a subset of Set D? Explain your answer.

5 Set C = {1, 2, 3, 4, 5, 6, 7, 8, 9, 10} and Set D = {2, 4, 6, 8, 10 …}

Write these sets.

a $C \cup D$ **b** C' **c** $C \cup D'$

24.2 Venn diagrams

Relationships between sets can be shown on a **Venn diagram**.

The universal set contains all the elements being considered and is shown as a rectangle.

Example 3	**Q** ξ = {1, 2, 3 … 12}
	A = {1, 3, 5, 7, 9, 11}
	Show the sets A + A′ on a Venn diagram. Note, A + A′ = ξ.

A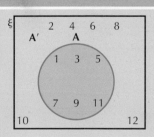

Example 4	**Q** **a** A = {3, 6, 9, 12, 15, 18}
	B = {2, 4, 6, 8, 10, 12, 14, 16, 18}.
	Draw a Venn diagram to show this information.
	b Desribe the numbers in Set A.
	c Describe the numbers in Set B.
	d Describe the numbers in Set A ∩ B.

(continued)

A a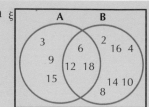

b The numbers in Set A are multiples of 3, less than 20.

c The numbers in Set B are multiples of 2, less than 20.

d Set A ∩ B = {6, 12, 18}. These are multiples of 6, less than 20.

Exam tips   Remember to include the rectangle in your Venn diagram.

Practice questions

1   Write the information in this Venn diagram using set notation.

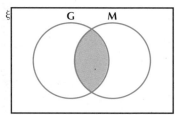

2   Which of these diagrams represents D ∪ E?

a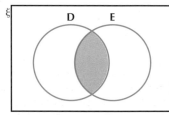

b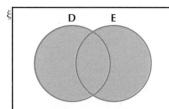

c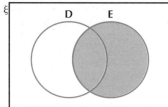

d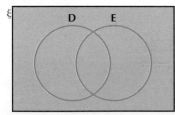

3   Set A = {red, blue, green} and Set B = {red, blue, green, yellow}

Represent this information on a Venn diagram.

4   Lucy has ten cards, numbered 1 to 10.

Draw a Venn diagram to show the two sets of 'odd numbers' and 'square numbers' for these cards.

5 Carlo has ten cards, numbered 1 to 10.

  a Draw a Venn diagram to show the two sets 'prime numbers' and 'factors of 10' for these cards.

  b Are the two events 'prime numbers' and 'factors of 10':

   i mutually exclusive   ii exhaustive?

   Explain your answers.

24.3 Using Venn diagrams to solve probability questions

You can use Venn diagrams to help solve questions about probability.

Example 5

Q Out of 40 students, 14 study French and 29 study German. 5 students study French and German.

  a Draw a Venn diagram to represent this information.

  b How many students study neither French nor German?

  c How many students study at least one of these languages?

  d What is the probability that a student picked at random studies German but not French?

A a

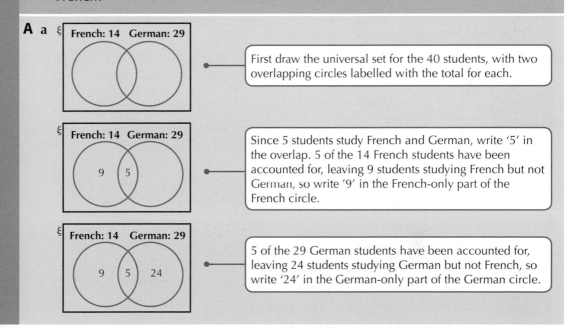

First draw the universal set for the 40 students, with two overlapping circles labelled with the total for each.

Since 5 students study French and German, write '5' in the overlap. 5 of the 14 French students have been accounted for, leaving 9 students studying French but not German, so write '9' in the French-only part of the French circle.

5 of the 29 German students have been accounted for, leaving 24 students studying German but not French, so write '24' in the German-only part of the German circle.

(continued)

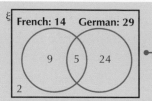

A French: 14 German: 29

9 5 24

2

This tells you that a total of 9 + 5 + 24 = 38 students are studying either French or German or both. This leaves 2 students unaccounted for, so they must be the ones taking neither class.

b Two students study neither German nor French.

c 38 students study at least one of these languages.

d There is a $\frac{24}{40} = 0.6 = 60\%$ probability that a student chosen at random from this group studies German but not French.

Example 6

Q The Venn diagram represents the subject choices of 70 students. Of these students, 8 students study only Psychology, 17 students study both Psychology and Sociology and 16 students study neither Psychology nor Sociology.

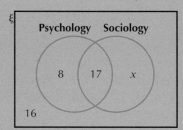

a Work out the value of x.

b How many students study Psychology?

c What is the probability that a student chosen at random does not study Psychology or Sociology?

A **a** $x = 70 - 8 - 17 - 16$

$x = 29$

b $17 + 8 = 25$ students study Psychology

c P(do not study Psychology or Sociology) $= \frac{16}{70}$

$= \frac{8}{35}$

Example 7

Q Poppy asked 50 people which types of chocolate they liked from plain (P), milk (M) and white (W). All 50 people liked at least one flavour.

19 people liked all three flavours.

16 people liked plain and milk, but not white.

21 people liked milk and white.

24 people liked plain and white.

40 people liked milk chocolate.

1 person liked only white chocolate.

a Draw a Venn diagram to represent this information.

b Poppy chose one of the 50 people at random. Work out the probability that the person liked plain chocolate.

c Given that the person selected at random liked plain chocolate, find the probability that this person liked exactly one other type of chocolate.

A a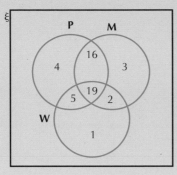

b 4 + 16 + 19 + 5 = 44

P(Plain chocolate) = $\frac{44}{50}$ = $\frac{22}{25}$

c 16 + 5 = 21

P(plain and one other type of chocolate) = $\frac{21}{44}$

Exam tips Drawing a Venn diagram to represent the given information can help you to answer a probability question.

Practice questions

1 How many elements are in Set A?

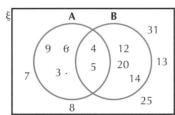

2 35 scuba divers were asked where they had dived.

22 of the 35 said the Irish Sea (I).

15 of the 35 said the Red Sea (R).

9 of the 35 said both the Irish Sea and the Red Sea.

a Draw a Venn diagram to represent this information.

b Work out the probability that a diver selected at random from this group has not dived in either the Irish Sea (I) or the Red Sea (R).

c Work out P(I ∪ R).

3 The Venn diagram shows the number of people in a park walking dogs (D) and the number wearing coats (C).

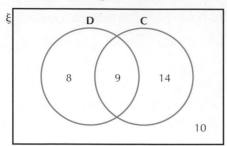

   a How many people are in the park?

   b What is the probability that someone chosen at random is wearing a coat?

   c Work out P(C′).

   d Work out P(D ∩ C).

   e Work out P(D ∪ C).

   f Work out the probability that someone in the park is walking a dog but not wearing a coat.

4 Kris makes and sells jewellery. Last week she made 42 pieces in silver. She made 21 necklaces, of which 11 were silver. 23 of her pieces were neither silver nor a necklace. Find the probability that a piece chosen at random from those she made last week is silver but not a necklace.

5 There is a bookcase for second-hand books in the seating area of the leisure centre. Of the 63 books, 49 are fiction and of those, 35 are paperback. In total, there are 46 paperback books.

   a What is the probability that a book chosen at random is a non-fiction paperback?

   b How many books are neither paperback nor fiction?

REVISION CHECKLIST

- Set A ∪ B contains elements belonging to A or B or both.
- Set A ∩ B contains elements belonging to both A and B.
- An empty or null set Ø contains no elements. So, for example, if A and B have no elements in common, then A ∩ B = Ø.
- A universal set contains all possible elements.
- Sets can be shown in a Venn diagram.
- Venn diagrams can help you to find the answers to probability questions.

Exam-style questions

1 Set R = {toffee, strawberry, vanilla, Turkish delight, cherry, banana, custard}

Which of these are subsets of A?

Set P = {strawberry}

Set Q = {cherry, banana and custard}

Set S = {strawberry, vanilla, Turkish delight, cherry, banana and custard, apple}

Set T = {toffee, cherry, banana and custard}

2 Set A = {dog, cat, mouse, guinea pig, hamster, tortoise}
Set B = {tiger, lion, panther, leopard, cheetah, cat}

Sets A and B contain all elements of the universal set.

a What is A ∪ B?

b Write out the elements of A′.

c List the elements of A ∩ B′.

3 The universal set = {1, 2, 3, 4, 5, 6, 7, 8, 9, 10} and Set P = {2, 3, 4, 5, 6}.

Write down the elements of P′.

4 Set A and Set B are in the universal set. Draw a Venn diagram to show:

a A ∪ B   **b** A ∩ B   **c** A′.

5 Set A and Set B are in the universal set. Draw a Venn diagram to show:

a A ∩ B   **b** A′ ∩ B′   **c** A ∪ B′.

6 Which of these diagrams represents D ∩ E?

a 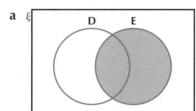   **b**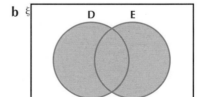

c 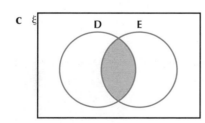   **d**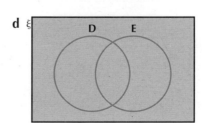

7 Set A = {even numbers} and Set B = {prime numbers}.

Draw a Venn diagram and write in the integers 11 to 20 inclusive.

8 Set X = {capital letters containing a straight line} and Set Y = {capital letters containing a curved line}.

Create a Venn diagram for the letters on these cards.

| R | E | S | U | L | T |

9 The table shows the results of a survey of about tablet ownership.

Represent this information using a Venn diagram.

	Did not own a tablet	Owned a tablet
Male	7	25
Female	8	10

10 Alice works in a hospital. The table shows some information about the patients in her care.

	Had a hip operation	Did not have a hip operation
Male	3	2
Female	20	5

a Represent this information in a Venn diagram.

b How many elements are in the universal set?

11 How many elements are in A ∪ B?

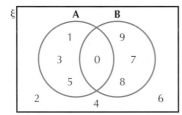

12 75 people queued to buy tickets to a show. Everyone in the queue managed to buy at least one ticket. 62 people bought a ticket for Horses Alive and 43 bought a ticket for Monster Wars.

Copy and complete the Venn diagram.

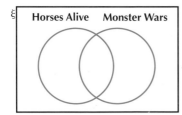

13 The Venn diagram shows the subjects studied by students in the library.

a How many of the students study Food Tech?

b How many of the students study neither Maths nor Food Tech?

c How many of the students are in the library?

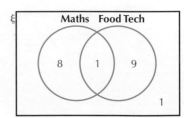

14 There are 15 people in a supermarket who can work on the tills or stack the shelves. Six can do both jobs and seven can only work on the till. What is the probability that a person chosen at random can only stack shelves?

15 Arthur sold 50 plants yesterday.

He sold 31 chilli plants, 14 of which were in trays.

Nine of his sales were neither chilli nor in a tray.

Find the probability that a sale that he made yesterday is a plant in a tray that was not a chilli.

16 All 150 members of an athletics club are either runners, field competitors or both.

112 are runners and 65 are field competitors.

a How many members are both runners and field competitors?

b What is the probability that a member, chosen at random, will be a field competitor only?

17 82 people were stopped as they came out of a supermarket.

In total, 65 people had bought food and 41 of these had also bought clothes.

Five had bought only clothes.

What is the probability that a person chosen at random had not bought food or clothing?

18 On Tuesdays, the leisure centre holds a yoga class followed by a circuit class.

Last week, 33 people attended the circuit class. Of these, 12 also attended the yoga class. A total of 50 people attended the two classes.

a What is the probability that a person chosen at random attended both classes? Give your answer as a fraction in its simplest form.

b How many people attended just the yoga class?

Now go back to the list of objectives at the start of this chapter.
How confident do you feel now about each of them?

25 Statistics

Go to 25.1

Objectives

Before you start this chapter, mark how confident you feel about each of the statements below:

	▶	▶▶	▶▶▶
I can recognise different types of data: primary, secondary, quantitative and qualitative.			
I can design a data collection sheet.			
I can identify and prevent bias in sampling.			
I can design, complete and interpret a frequency table.			
I can design, complete and use two-way tables.			
I can draw and interpret pictograms and line graphs, including vertical line charts.			
I can draw, interpret and compare different types of bar charts (composite, comparative).			
I can draw, interpret and compare pie charts.			
I can draw and interpret frequency diagrams (histograms) for grouped continuous data of equal class widths.			
I understand when to use a pie chart.			

Check-in questions

- Complete these questions to assess how much you remember about each topic. Then mark your work using the answers at the back of the book.

- If you score well on all sections, you can go straight to the Revision checklist and Exam-style questions at the end of the chapter. If you don't score well, go to the chapter section indicated and work through the examples and practice questions there.

1 Which of these are primary data and which are secondary?

Go to 25.1

  a Finding out information about a holiday destination by looking on the internet.

  b Measuring the height of each student in your class.

  c Finding out the shoe size of each student in your class.

  d Looking at records to see how many babies were born in January.

  e Looking at tables about the number of road traffic accidents each year.

2 Explain why this sample is biased. Go to 25.1

'Investigating the pattern of absences for a college by studying the registers in February.'

3 A market research company interviewed people travelling by car and by train.

45 out of the 100 car travellers had travelled 20 miles or fewer.

Of the 250 people interviewed, 85 had travelled more than 20 miles.

Design a two-way table and work out how many people had travelled more than 20 miles by train. Go to 25.2

4 The pictogram shows the number of hours of sunshine on one day in August.

Hours of sunshine

Manchester	☀ ☀ ☀ ☀
Cairo	☀ ☀ ☀ ☀ ☀
Brisbane	☀ ☀ ☀

Key ☀ = 2 hours

a Which city had the fewest hours of sunshine?

b What is the difference between the hours of sunshine recorded for Cairo and Manchester? Go to 25.3

5 Colin investigated the colours of front doors on a housing estate. There were only five different colours.

Together, the frequency table and bar chart show his results.

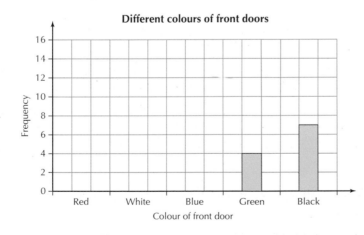

Different colours of front doors

Colour	Frequency
Red	15
White	8
Blue	10

a Copy and complete the bar chart.

b Work out the total number of houses in Colin's investigation. Go to 25.3

6 Daisy owns a flower shop. The bar chart shows the percentages of each type of flower she has sold for the last two years.

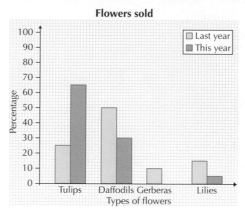

Flowers sold

a Write down the most popular flower sold:
  i this year   **ii** last year.

b What percentage of the total flowers sold were lilies:
  i this year   **ii** last year?

Go to 25.3

7 Draw a pie chart for this data.

Go to 25.4

Favourite colour	Frequency
Blue	15
Red	9
Black	5
Green	7

8 The pie chart shows information about some students' favourite football teams.

15 students chose Liverpool as their favourite team.

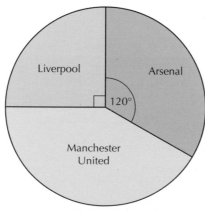

Favourite football teams

a How many students chose Arsenal as their favourite team?

b How many students chose Manchester United as their favourite team?

Go to 25.4

25.1 Collecting data

Data is a collection of facts and information, such as numbers or measurements. Data is often collected to investigate or test a **hypothesis**, which is a statement that might or might not be true.

Types of data

- Data is either **qualitative** or **quantitative**.

  Quantitative data is numerical. For example, the number of cars in a car park.

  Qualitative data is not numerical. For example, the colours of cars in a car park. (Quantitative data can be discrete or continuous.)

- Data is either **discrete** or **continuous**.

  Discrete data can only take particular values; it is often found by counting. For example, the number of cars in a car park.

  Continuous data can take any value in a given range; it is often found by measuring. For example, the height and weight of an adult.

- Data is either **primary data** or **secondary data.**

  Primary data is collected by the person who is going to analyse and use it.

  Secondary data has been collected by someone else, for example, through a census. A census is carried out in Britain every 10 years. The data is analysed by the authorities so that they can make plans for the future.

Collecting data

A **data collection sheet** is used to record data.

The data collection sheet below has been designed to collect data to test the hypothesis 'Most college staff drive black cars'.

Colour of staff cars

Colour	Tally	Frequency
Red		
Blue		
White		
Green		
Black		
Other		

Before collecting data, consider these questions.

- Is the data collection sheet clear and easy to use?
- Will the data recorded on the data collection sheet enable you to test the hypothesis?
- How much data is needed?
- Will the time and place of the data collection affect the results?

You can also conduct an experiment to test a hypothesis, for example: *The better the light, the faster seedlings will grow.*

Variable: the intensity of the light

Conditions: All other conditions must stay the same. At the start of the experiment, all the seedlings must be as uniform as possible (size, strength and colour).

Possible sources of bias meaning the experiment needs to be started from scratch: one side of the tray gets more sunlight.

When conducting an experiment, consider these questions.

• Does the experiment test the hypothesis?

• How many results are needed to be sure of an accurate representation?

Sampling

It is not always possible or practical to ask or test an entire **population**, so **sampling** is used. It is important that the sample is representative of the population and does not contain bias. The bigger the sample, the more accurate the results will be.

In a **random sample**, each member of the population has an equal chance of being chosen. For example, when the names of all the students in a class are put in a hat, each student has an equal chance of being chosen.

Practice questions

1   Which of these is a hypothesis?

   a Who do you think can run for longer, men or women?

   b Can men run for longer than women?

   c Men can run for longer than women.

   d I'm going to prove men can run for longer than women.

2   Copy and complete the table and tick the box(es) that apply to each data set.

	Qualitative	Quantitative	Discrete	Continuous
Colour of car				
Make of car				
Engine size				
Number of seats				
Top speed				

3   Ben wants to find out the best-selling car of 2015. Should he use primary or secondary data? Give a reason for your answer.

4   Design a data collection sheet for a survey to find out how many books people borrow from the college library each week. Explain how you can ensure your sample is unbiased.

5   Explain how to take a random sample from 250 passengers on a train.

25.2 Organising data

Grouped frequency tables

If a large amount of numerical data are collected, it is usual to group it in class intervals.

Each class interval is usually the same width, but will depend on the range of the data. Intervals of 2, 5 or 10 are common. The class intervals must not overlap.

For example, test scores out of 50 (discrete data) might be grouped as: 0–10, 11–20, 21–30, 31–40 and 41–50. Note that the first group has a range of 11 and the others have a range of 10.

Score	Tally	Frequency
0–10	|||	3
11–20	|||| ||	7
21–30	||||	4
31–40	|||| ||||	10
41–50			1

For continuous data, the class intervals are usually written using inequalities. For example, the table below shows the heights in centimetres of 30 members of a youth club.

Height (cm)	Tally	Frequency
$120 \le h < 130$	||||		6
$130 \le h < 140$	||||	5
$140 \le h < 150$	|||| |||| ||||	14
$150 \le h < 160$	||||	5

$120 \le h < 130$ means that the heights are greater than or equal to 120 cm and less than 130 cm. A height of 130 cm would be counted in the interval $130 \le h < 140$.

Two-way tables

A two-way table can be used to show the frequencies of different variables. They are often found in probability questions. (See Chapter 23 for more on probability.)

For example, this two-way table shows the results of a survey to find out the favourite genre of films for a group of Year 13 students.

	Action	Comedy	Science fiction	Total
Male	20	10	15	45
Female	30	20	10	60
Total	50	30	25	105

The table shows that 20 males chose action films as their favourite genre and 30 students chose comedies.

> **Exam tips** Always check that the total of the rows and the total of the columns are equal.

Practice questions

1   Sam grows tomato plants to sell at a fair. His plants range from 2 weeks to 6 weeks in growth. He has a different number of plants at each stage. Design a frequency table that Sam can complete to show the numbers of plants he has at each stage of growth.

2   The frequency table shows the heights of children in a clinic. Make one criticism of the frequency table and suggest how to improve it.

Height (cm)	Tally	Frequency
80–90		
90–100		
100–110		
110–120		
120–130		

3   An online app store asked users about their favourite music genre. The results are shown in the two-way table.

	Rock	Pop	Classical	Country
Male	263	148	23	0
Female	134	204	19	4

  a   How many people were asked in total?

  b   How many people chose rock as their favourite music genre?

  c   How many more males than females prefer classical music?

4   The table shows some information about how many tablets and phones there are in households. Copy and complete the table.

	0	1	2	3	4	5+	Total
Tablets	3	6	21		18		78
Phones		11	13	44	69	22	
Total	4			68			

5   The points scored in football matches are often presented in a two-way table. Research the current top-ten teams in Division One and produce a two-way table to show their points.

25.3 Displaying data

Once data has been collected, it can be displayed using a range of diagrams.

The type of diagram that you use depends on the type of data that you have to display.

Pictograms

A pictogram is often used to represent qualitative data. Each picture represents an item or a number of items.

Choose a simple symbol that can be divided if necessary and include a key.

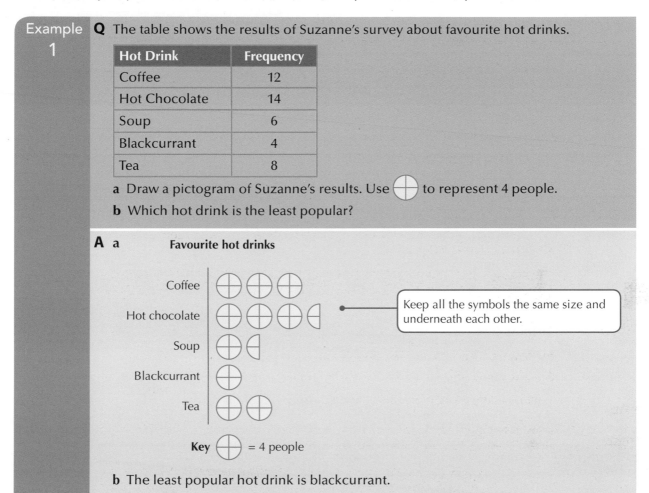

| Example 1 | **Q** The table shows the results of Suzanne's survey about favourite hot drinks. |

Hot Drink	Frequency
Coffee	12
Hot Chocolate	14
Soup	6
Blackcurrant	4
Tea	8

a Draw a pictogram of Suzanne's results. Use ⊕ to represent 4 people.

b Which hot drink is the least popular?

A a　　　**Favourite hot drinks**

Coffee

Hot chocolate

Soup

Blackcurrant

Tea

Keep all the symbols the same size and underneath each other.

Key ⊕ = 4 people

b The least popular hot drink is blackcurrant.

Bar charts and vertical line charts

Bar charts and vertical line charts are useful for showing patterns or trends in discrete data.

The bars can be vertical or horizontal, but they must be of equal width and have a gap between each one.

When you draw a bar chart, choose a sensible scale and remember to label the axes.

| Example 2 | **Q** The frequency table shows the types of films that Bina's friends liked best. Draw a bar chart for this data. |

Types of films	Frequency
Drama	5
Science fiction	3
Thriller	1
Comedy	6
Horror	2

(continued)

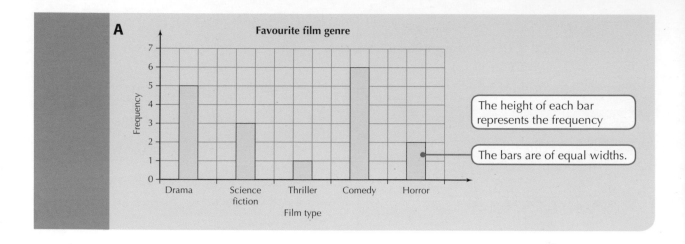

A line graph can be used to show discrete data. It is drawn in the same way as a bar chart but uses a thick line instead of a bar.

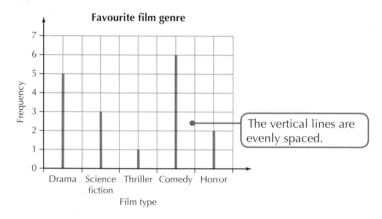

The vertical lines are evenly spaced.

Exam tips Remember to label the axes on the diagram.

Comparative and composite bar charts

Comparative bar charts can be used to compare data. In a comparative bar chart, two (or more) bars are drawn side-by-side.

Example 3

Q Samuel carried out a survey about favourite sports. The table shows his results.

Favourite sport	Number of males	Number of females
Swimming	6	8
Football	15	3
Hockey	5	10
Tennis	9	14

Draw a comparative bar chart to show his data.

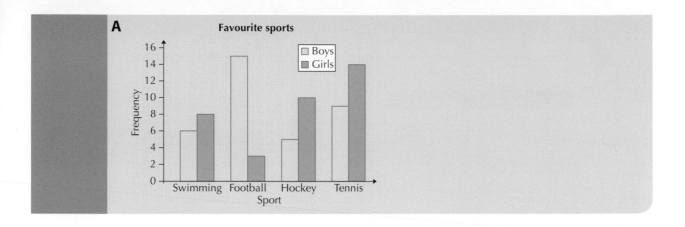

A

Favourite sports

You can also use a composite bar chart to show and compare two or more sets of data.

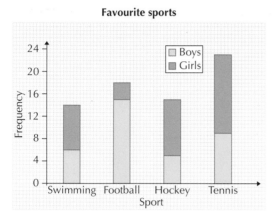

Favourite sports

Both bar charts show that considerably more males than females like football and more females than males like tennis.

Frequency diagrams for continuous data

If the data is continuous, the bars must touch. This is known as a frequency diagram or histogram. The data must be grouped into equal class intervals if the length of the bar is used to represent the frequency.

For example, the masses of 30 workers in a factory are shown in the table.

Remember the following:

- The axes do not need to start at zero. Attention is usually drawn to this fact by using a jagged line like this:
- Frequency always goes on the vertical axis.
- For continuous data there are no gaps between the bars.
- The width of each bar is the same as the class interval.
- The axes are labelled and the graph has a title.

The diagram shows that the largest group of workers has a mass between 55 and 65 kg.

Mass (kg)	Frequency
$45 < M \le 55$	7
$55 < M \le 65$	13
$65 < M \le 75$	6
$75 < M \le 85$	4
Total	30

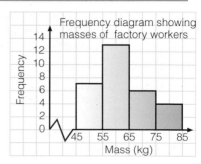

Frequency diagram showing masses of factory workers

1 Necati is opening a snack bar near the local college. In order to decide what to sell, he carried out a survey about favourite snacks. His results are shown in the table.

Type of snack	Frequency
Crisps	12
Biscuits	8
Fruit	15
Chocolate	21
Sweets	7
Sandwiches	11

a How many people did he ask altogether?

b How many people said fruit was their favourite snack?

c Draw a pictogram to represent this information. Remember to include a key.

2 The table shows the numbers of cars sold at a garage in one week.

Day	Monday	Tuesday	Wednesday	Thursday	Friday
Number of cars sold	9	10	11	10	12

Sam uses this information to draw a bar chart.

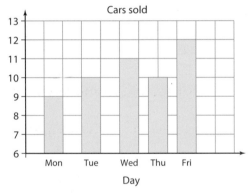

a A customer looks at the bar chart and says 'Twice as many cars were sold on Friday as on Monday.' Is the customer correct? Show how you decide.

b Another customer points out some errors in the bar chart. Write down two mistakes that Sam has made.

3 Erin conducted a survey about the favourite genres of fiction among members of her book group.

Type of fiction	Frequency
Romance	7
Thriller	2
Mystery	10
Sci-fi/fantasy	4
Historical	1

Represent this data in a vertical line graph.

4 List the advantages and disadvantages of different methods of representing data. Consider pictograms, bar charts, comparative bar charts and vertical line graphs.

5 Kaz and Bill go to a quiz night. There are eight rounds in the quiz. The table shows the number of points they each scored in the each round. Draw a comparative bar chart to show this data.

	Round							
	1	2	3	4	5	6	7	8
Kaz	10	6	3	7	6	9	20	4
Bill	9	7	5	6	0	20	4	7

6 The heights of some athletes were measured.

Height (cm)	Frequency
$150 \leq h < 155$	5
$155 \leq h < 160$	8
$160 \leq h < 165$	11
$165 \leq h < 170$	7
$170 \leq h < 175$	3

Draw a frequency diagram to show this data.

25.4 Pie charts

Pie charts are another method of displaying data. They are circles split up into sectors. Each sector represents a set of data.

Pie charts are particularly useful for comparing proportions, but can be misleading as the actual values are not shown.

Drawing a pie chart

When you draw a pie chart:

1 Calculate the angles.
  - Find the total number of items in the data.
  - Work out the number of degrees that will represent one item (360 ÷ total number of items).
  - Work out the size of the angle that will represent each category (multiply the number of degrees that represent one item by the number of items in the category).

2 Draw the pie chart accurately. Your angles must be accurate, within 2°.

3 Label the sectors.

Example
4

Q The table shows the favourite subjects of 48 college students. Draw a pie chart to show this information.

Subject	Frequency
Maths	18
English	8
Art	10
Geography	12
	48

A There are 48 students, so $\frac{360°}{48} = 7.5°$ degrees represents 1 student.

Multiply each frequency by 7.5° to find the angles for each subject.

Subject	Frequency	Angle on pie chart
Maths	18	135°
English	8	60°
Art	10	75°
Geography	12	90°
	48	360°

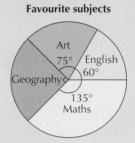

Favourite subjects

Draw the pie chart and label each sector.

Exam tips   When drawing a pie chart remember to label each sector and check that you have measured the angles accurately. They must be within + or –2 degrees.

Interpreting a pie chart

You will need to measure the angles in a pie chart carefully if they are not given.

Example
5

Q The pie chart shows the favourite sports of 18 college students.
How many students like:

a tennis   **b** football   **c** hockey?

A 18 students are represented by 360°, so 1 student is represented by 360° ÷ 18 = 20°.

120° ÷ 20 = 6 students like tennis

160° ÷ 20 = 8 students like football

80° ÷ 20 = 4 students like hockey

Note that 6 + 8 + 4 = 18, the total number of students. This is a useful check.

Comparing pie charts

When comparing data presented in pie charts, it is important to remember that sectors that are the same size on two or more pie charts are unlikely to represent the same number of items.

Example 6

Q The two pie charts show the sales of different flavours of ice cream in two shops on the same day.

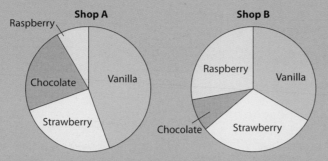

Thomas says: 'More people bought vanilla ice cream from Shop A than Shop B.'

Explain why Thomas might be wrong and rewrite his statement so that it is definitely true.

A Thomas might be wrong because the pie charts do not tell us how many people bought ice cream from either shop. Although the sector for sales of vanilla ice cream is larger in shop A, it could represent fewer people.

He could say that proportion of vanilla ice cream sales was higher in shop A than in shop B.

> **Exam tips** When comparing pie charts, use data and calculations to support your explanation. Remember that unless you are given the totals you can only compare proportions.

Practice questions

1 Jake wants to draw a pie chart to show the numbers of hours he spends on different activities each week.

Copy and complete his table.

Activity	Number of hours	Angle on a pie chart
Reading	5	
Cycling	15	
Gym	6	
College	24	
Work	10	

2 The table shows the favourite lecturers of 45 college students. Draw a pie chart to represent the data.

Favourite lecturer	Frequency
Mr Fitz	22
Mr Barton	10
Mrs Bell	4
Miss Combwell	6
Miss Dhillon	3

3 A pet-food manufacturer surveyed some of its customers about the number of dogs they own. The pie chart shows the results.

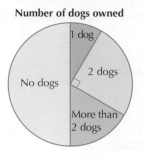

Number of dogs owned

   a What percentage of those surveyed owned two dogs?

   b Compare the number of people who did not own a dog with those that did own a dog.

   c What fraction of the people surveyed owned more than two dogs?

   d 200 people said they owned one dog. How many people completed the survey?

4 A holiday company surveyed customers about the type of accommodation they had. The pie charts show the data collected in summer and winter.

Oshra says that more people stayed in B&B in summer than in winter. Is she correct? Explain your answer.

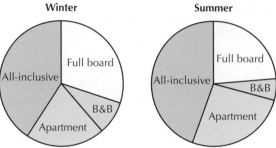

5 Davey is writing a report about the different types of games his friends like. He surveys his online friends and his college friends separately and plans to draw a pie chart to show each set of data. However, there are a different number of people in the two groups. When he draws the pie charts side-by-side, how can he alert people to the fact that the two pie charts do not represent the same number of people? Why is it important for him to do this?

REVISION CHECKLIST

- In a survey, a sample is taken to represent the population. A sample that is too small or is not representative can mean the results are biased.

- In a random sample, each member of the population has an equal chance of being chosen.

- A pictogram can be used to represent qualitative data. Each picture represents an item or a number of items. A pictogram must include a key.

- A comparative bar chart has two or more bars drawn side-by-side. It can be used to compare data.

- If the data is continuous, the bars must touch. This is known as a frequency diagram or histogram. The data must be grouped into equal class intervals if the length of the bar is used to represent the frequency.

- To draw a pie chart, first work out how many degrees will represent one item and then multiply by the number of items in the category to find the angles. Finally, draw the angles accurately and label the sectors.

- To read or interpret a pie chart, first work out what 1° or one item is worth and then use this unitary amount to find the number of items in each category.

- Comparative bar charts are useful when comparing two or more sets of discrete data, although they can be misleading if the axes do not start at zero. Pie charts are particularly useful when comparing proportions, but can be misleading when the actual values are not shown.

Exam-style questions

1 Mr Beech teaches violin and piano. He wants to know if learning both instruments means you get higher scores in exams. Write a suitable hypothesis for him.

2 Terry grows vegetables to sell. He can use either Fertilizer A or Fertilizer B. He is going to conduct an experiment to see which fertilizer is better. Write a suitable hypothesis for him.

3 **a** List two advantages of using primary data.

   b List two advantages of using secondary data.

4 Jodie wants to know how many people use the gym in her village. She designs a survey and asks all the people that are in the gym on Friday night to complete it. Explain why this is not a good method of data collection.

5 Marcus owns 42 hats. He has 15 baseball caps, of which 9 are red. He has 11 other red hats. How many hats does he have that are neither red nor a baseball cap?

6 Eddie plays for Longfield United FC. The list below shows the number of goals the team have scored over the last twenty matches.

0, 5, 1, 4, 0, 0, 0, 3, 1, 3, 0, 4, 1, 1, 1, 1, 2, 3, 0, 0

Copy and complete the frequency table to show this data.

Number of goals	Tally	Frequency
0		
1		
2		
3		
4		
5		

7 Copy the table. Tick the appropriate columns to show whether these data sets are qualitative, quantitative, discrete or continuous. You may tick more than one box for each data set.

Data set	Qualitative	Quantitative	Discrete	Continuous
Length of a TV programme				
Colour of a shirt				
Dress size				
Make of car				

8 Janice plans to open a café in town. She wants to find out people's favourite hot drinks and snacks. She needs to decide between asking the entire population or asking a sample of the population.

   a Give one advantage of asking the entire population.

   b Give one advantage of asking a sample of the population.

9 The table shows how shoppers travelled to a supermarket.

Transport	Bus	Car	Train	Walk	Cycle
Frequency	12	44	8	17	21

Draw a bar chart to show this information.

10 The pictogram shows the results of asking 40 people which takeaway they prefer.

a How many people prefer fish and chips?

b How many more people prefer pizza than indian?

c Colin says that more than three-quarters of the people chose either pizza or fish and chips. Is he correct? Give a reason for your answer.

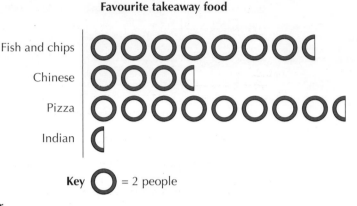

Favourite takeaway food

Key ◯ = 2 people

11 Roseanna made a note of how many minutes she spent on the internet during one week. Her results are shown below. Draw a vertical line graph to show this information.

Monday   2 hours

Tuesday   30 minutes

Wednesday   1 hour

Thursday   80 minutes

Friday   20 minutes

Saturday   90 minutes

Sunday   140 minutes

12 The table shows some information about Jasmine's book collection.

Genre	Hardback	Softback	E-book
Non-fiction	12	3	0
Mystery	0	23	55
Romance	2	5	4
Fantasy	18	34	28
Historical	0	1	5
Children's	19	21	0

a How many hardback books does she own?

b How many mystery books does she have as e-books?

c How many fantasy books does she own in total?

d What is the probability that a book, chosen at random, is an e-book?

13 The bar chart shows the results of Mo asking people to choose their favourite form of digital communication.

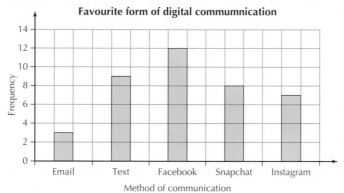

a Which method of communication was most popular?

b How many people did she ask altogether?

c How many more people preferred Facebook to email?

14 The pie chart shows which colour pen people are using to write their lecture notes. There are 24 people in the lecture.

a How many of them are using a pen that is not blue or black?

b What is the probability that a person chosen at random will be using a black pen? Give your answer as a fraction in its simplest form.

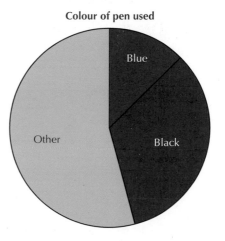

15 The pictogram shows how many tubs of ice cream were used in a shop last week.

Use this information to draw a pie chart.

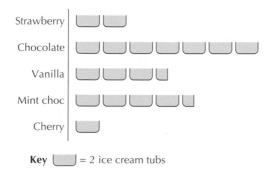

16 These two pie charts show the colours of car preferred by two groups of people. Pie chart A represents 200 people and pie chart B represents 50 people.

Which of these statements must be true?

Colour of car preferred by group A

Colour of car preferred by group B

a More people in Group B than Group A prefer blue cars.

b More than half the people in Group B prefer black cars.

c Twice as many people from Group A as Group B prefer blue cars.

17 Explain the difference between a composite bar chart and a comparative bar chart.

You may use diagrams to support your explanation.

18 Explain why Jessica should not use a pie chart to represent the information in the table.

Type of pet	Cat	Dog	Hamster	Horse
Number of males who own this pet	14	22	7	1
Number of females who own this pet	22	17	0	8

19 The table shows the different cars represented in a supermarket car park.

Car manufacturer	Nissan	Ford	Vauxhall	Audi	BMW	Skoda	Other
Frequency	81	47	21	5	2	33	140

a Draw a pie chart to represent this information.

b What is the probability that a car chosen at random is a Skoda? Give your answer correct to two decimal places.

Now go back to the list of objectives at the start of this chapter. How confident do you now feel about each of them?

26 Averages

Check-in questions

- Complete these questions to assess how much you remember about each topic. Then mark your work using the answers at the back of the book.

- If you score well on all sections, you can go straight to the Revision checklist and Exam-style questions at the end of the chapter. If you don't score well, go to the chapter section indicated and work through the examples and practice questions there.

1 Work out the mean and range for each set of data. **Go to 26.1**

  a 2, 7, 9, 3, 6, 4, 5, 2   **b** 7, 9, 11, 15, 2, 1, 6, 12, 19, 13

2 Find the median and mode for each set of data. **Go to 26.1**

  a 2, 7, 1, 4, 2, 2, 3, 7, 9, 2   **b** 6, 9, 11, 11, 13, 6, 9, 6, 6, 4, 1

3 Reece created this table for the number of minutes (to the nearest minute) it took some students to complete a maths problem.

Number of minutes to solve problem	5	6	7	8	9	10
Frequency	4	7	10	4	3	1

  Work out: **a** the mean   **b** the median   **c** the mode.

  d Rupinder says, 'The mean time taken to solve the problem is 12 minutes.' Explain why Rupinder cannot be correct. **Go to 26.2**

4 The table shows the heights, h cm, of some students.

Height, h (cm)	Frequency, f	Midpoint, x	fx
$140 \leq h < 145$	4		
$145 \leq h < 150$	9		
$150 \leq h < 155$	15		
$155 \leq h < 160$	6		

Work out an estimate for the mean height of the students. **Go to 26.3**

5 **a** Draw an ordered stem and leaf diagram for this data.

27	28	36	42	50	18
25	31	39	25	49	31
33	27	37	25	47	40
7	31	26	36	9	42

b Work out the median. **Go to 26.4**

26.1 Comparing data: averages and spread

An **average** gives you an idea of a 'typical' value for a set of data.

You should understand and be able to calculate the mean, the median, the mode and the range.

- The **mean** is the most commonly used average: $\text{mean} = \dfrac{\text{sum of a set of values}}{\text{the number of values used in the set}}$

  For example, the mean of 1, 2, 3, 3, 1 is $\dfrac{1+2+3+3+1}{5} = 2$.

- The **median** is the middle value of an ordered data set.

  For example: the median of 2, 2, 3, 3, 7, 9, 11 is 3;
  the median of 1, 2, 3, 4, 5, 6 is halfway between 3 and 4, so 3.5.

- The **mode** is the value in a data set that occurs the most often.

  For example, the mode of 2, 2, 2, 3, 5, 7 is 2.

- The **range** shows the spread of a set of data. It can be calculated using:
  range = highest value – lowest value.

  It is not an average, but useful for comparing sets of data.

  For example, the range of 1, 2, 3, 4, 7, 10 is 10 – 1 = 9.

The table below gives some information about when to use each type of average.

Average	Advantages	Disadvantages
Mean	Uses all of the numbers in the data set.	Affected by extreme values (outliers). Actual value may not exist.
Median	Not affected by extreme values (outliers).	Actual value may not exist.
Mode	Not affected by extreme values (outliers) Can be used with qualitative data.	May not exist or there may be more than one mode, particularly if the data set is small.

You can use averages, such as the mean or median, and measures of spread, such as the range, to compare two sets of data. First calculate the same average and range for both data sets, then write a sentence for each statistic comparing the values for each data set. Remember you can only make a valid statement if you can support it with statistical evidence.

Practice questions

1 These are the salaries for all seven employees of a small company.

   £8000   £18 000   £18 000   £24 000   £24 000   £25 000   £37 000

   a Work out the range.

   b Write down the modal salary.

   c Find the median salary.

   d Work out the mean salary.

2 Trisha works in a small library that is under threat of closure. She keeps a log of how many customers use the library each day. The figures for the last two weeks are:

   23   14   19   37   45   79   18   29   17   42   30   41

   She writes a report to support keeping the library open and wants to quote an average number of daily visitors.

   She says that the modal number of people using the library each day is 79.

   Explain why she is wrong.

3 Joachim is conducting research for a new medicine. He needs to collect data about the patients in the trial. He starts by collecting their heights, in centimetres, as shown below.

   147   153   154   160   178   181   184   190   201   202

   a Work out the range.

   b Write down the modal height.

   c Work out the median.

   d Work out the mean.

4 The mean mass of five basketball players is 68.4 kg. The mean mass of the five basketball players and their coach is 71 kg.

   What is the mass of the coach?

5 Ash runs a market stall selling dresses. He sold nine dresses in the first hour. The colours were: red, green, blue, red, red, blue, white, black, red.

   He says he cannot give the average of these as there are no numbers. Is he correct? Explain your answer.

26.2 Finding averages from a frequency table

You can also find the mean and spread of data presented in a frequency table.

To find the mean from a frequency table, use the formula: mean $(\bar{x}) = \dfrac{\Sigma fx}{\Sigma f}$, where

Σ means the sum of

f represents the frequency

$\bar{x}$ represents the mean.

Example 1

Q The table shows the shoe sizes of a group of dancers.

Shoe size (x)	3	4	5	6	7
Frequency (f)	5	18	22	15	5

a Work out the mean, range and median.

b John says, 'The mode is 22'. Explain why John is wrong.

A **a Mean** = $(\bar{x}) = \dfrac{\Sigma fx}{\Sigma f}$

Calculate fx, the sum of fx and the total of the frequencies in turn. It is good practice to add an extra column/row to the table for these values.

Shoe size (x)	3	4	5	6	7	
Frequency (f)	5	18	22	15	5	Total: 65
fx	15	72	110	90	35	Total: 322

Then divide the sum of fx by the sum of f: $\dfrac{322}{65}$ = **4.95 (2 d.p.)**

Range = highest value – lowest value

$\qquad$ = 7 – 3

$\qquad$ = **4**

There are 65 items of data. The **median** value is the 33rd value, since there are 32 sets of data on either side. This is within size 5, so the median shoe size is **5**.

b The modal shoe size is 5 (highest frequency). John has used the highest frequency itself rather than the shoe size.

Exam tips When you calculate the mean, be sure to divide by the total of the frequencies not the number of different shoe sizes.

Exam tips The position of the median in a set of n data values is $\dfrac{(n+1)}{2}$.

For an odd number of data values, there is one value in the middle.

For example, for 65 data values $\dfrac{(65+1)}{2}$ = 33.

For an even number of data values, there are two values in the middle.

For example, for 44 data values $\dfrac{(44+1)}{2}$ = 22.5 so the 22nd and 23rd are

the two values in the middle. The mean of these two values is the median.

Practice questions

1 Look at the data shown in the frequency table.

  a Work out the range.

  b Write down the mode.

  c Work out the median.

  d Work out the mean.

x	Frequency
2	7
3	3
4	5
5	6

2 Look at the data shown in the frequency table.

  a Work out the range.

  b Write down the mode.

  c Work out the median.

  d Work out the mean.

Shoe size	3	4	5	6	7	8	9
Frequency	4	9	12	19	10	4	2

3 Children in a hospital ward are given merits as rewards for taking their medicine, not disturbing others at night, going to the playroom, etc. This table shows the numbers of merits awarded to the patients over three months.

Number of merits	January	February	March
0	22	8	14
1	15	19	32
2	30	21	20
3	26	41	16
4	14	24	33
5	3	7	4

  a How many patients were there in January?

  b Work out the mean number of merits per patient in February. Give your answer correct to one decimal place.

4 Look at the data shown in the frequency table.

  a Work out the range.

  b Write down the mode.

  c Work out the median.

  d Work out the mean.

Dress size	6	8	10	12	14	16	18
Frequency	1	19	12	19	7	4	2

5 The numbers of assignments given to some college students are listed in the frequency table. The frequency in the last row is missing.

The median number of assignments given is 2.5.

Work out the mean number of assignments given. Give your answer correct to two decimal places.

Assignments given	Frequency
0	6
1	3
2	8
3	7
4	

26.3 Averages of grouped data

When data is grouped into **class intervals**, the exact values are not known so it is not possible to work out the actual mean or mode.

You can work out an **estimate for the mean** using the midpoints of the class intervals.

$\bar{x} = \dfrac{\sum fx}{\sum f}$ where $\bar{x}$ represents an estimate for the mean, f represents the frequency and x represents the midpoint of the class interval.

The **modal class** is the class interval with the highest frequency.

The **class interval containing the median** is found in the same way as for discrete data.

Example 2

Q Look at this data set.

Mass M (kg)	Frequency (f)
$30 \leq M < 35$	6
$35 \leq M < 40$	14
$40 \leq M < 45$	22
$45 \leq M < 50$	18

a Work out an estimate for the mean.

b State the modal class.

c Find the class interval that contains the median.

A **a** $\bar{x} = \dfrac{\sum fx}{\sum f}$

Add extra rows/columns to the table for values of the midpoint, x and the calculation fx.

Mass M (kg)	Frequency (f)	Midpoint (x)	fx
$30 \leq M < 35$	6	32.5	195
$35 \leq M < 40$	14	37.5	525
$40 \leq M < 45$	22	42.5	935
$45 \leq M < 50$	18	47.5	855
	Total: 60		**Total: 2510**

$\bar{x} = \dfrac{2510}{60}$

$\bar{x} = 41.8\dot{3}$

b The modal class is $40 \leq M < 45$.

c The median value is the $\dfrac{60+1}{2}$ value. In this case, the median lies halfway between the 30th and 31st values. These are both in the class interval $40 \leq M < 45$.

Hence, the median lies in the class interval $40 \leq M < 45$.

Exam tips | Adding two extra columns to the table, i.e midpoint (x) and fx, is helpful when working out the estimate of the mean.

Practice questions

1 Look at the grouped frequency table.

x	$0 < x \le 5$	$5 < x \le 10$	$10 < x \le 15$	$15 < x \le 20$
Frequency	16	27	19	13

   a Write down the modal class interval.

   b Work out an estimate for the mean.

2 Look at the grouped frequency table.

x	Frequency
$0 < x \le 10$	4
$10 < x \le 20$	9
$20 < x \le 30$	17
$30 < x \le 40$	13
$40 < x \le 50$	7

   a Write down the modal class interval.

   b Work out an estimate for the mean.

   c It is not possible to calculate the range for this data. Why not?

3 The table shows the heights of 70 seedlings.

Height, h (cm)	Frequency
$0 \le h \le 10$	4
$10 < h \le 20$	15
$20 < h \le 30$	30
$30 < h \le 40$	18
$40 < h \le 50$	3

   a Write down the modal class.

   b Work out an estimate for the mean.

   c Find the class interval that contains the median.

4 When data is presented in a grouped frequency table, you are never asked to work out the mean, only an estimate for the mean. Why is this?

5 When finding an estimate for the mean for data in a grouped frequency table, why is the midpoint of the class interval used rather than the minimum or maximum value?

26.4 Stem and leaf diagrams

Stem and leaf diagrams are useful for recording and displaying data. It is straightforward to find the mode, median and range of a set of data presented in this way.

Outliers are values that 'lie outside' most of the other values of a set of data.

For example, in the data set

     1, 1, 2, 2, 2, 3, 4, 4, 5, 6, 22

22 is the outlier.

Example 3

Q These are the marks gained by some college students in a maths test:

| 52 | 45 | 63 | 67 | 75 | 57 | 68 | 67 | 60 | 59 | 67 |

a Draw an ordered stem and leaf diagram to show this information.

b Work out the median.

c Write down the mode.

d Work out the range.

A a

4	5
5	2 7 9
6	0 3 7 7 7 8
7	5

Key   6 | 3   = 63 marks

b The median is the sixth value, so 63 marks.

c The mode is 67 marks.

d The range (highest value – lowest value) is 75 – 45 = 30 marks.

Back-to-back stem and leaf diagrams are very useful for comparing two sets of data.

Example 4

Q The same college students in Example 3 sat a second maths test. The results are shown in the back-to-back stem and leaf diagram.

Compare the two sets of data.

	Test 2		Test 1
	9 6 2	4	5
9 7 5 1 1		5	2 7 9
	3 1 0	6	0 3 7 7 7 8
		7	5

Key   Test 1   5 | 2   = 52 marks

Test 2   1 | 5   = 51 marks

A The median score in test 2 (55 marks) is lower than the median score in test 1 (63 marks). The range of the scores in test 2 (21 marks) is lower than the range of the scores in test 1 (30 marks). On average, the students did better in test 1 but their scores were less consistent than in test 2.

Exam tips   Always remember to include a key when drawing a stem and leaf diagram.

Practice questions

1 Alice asks everyone in the canteen at work to count the total value of the coins in their purse or wallet. The amounts are shown below:

| £4.22 | £3.16 | 18p | 0 | £5.69 | £1.63 | £2.03 | 67p | £3.84 | 79p | £2.34 |

Draw an ordered stem and leaf diagram to show this information.

2 Explain what is meant by the term 'outlier'.

3 As part of an annual health check, the heights of all visitors to a Senior Citizens' Day Centre are measured. The heights of the 27 visitors, to the nearest centimetre, are given below.

157	165	184	179	148	155	177
155	149	164	149	150	132	180
166	141	157	160	164	157	178
178	187	195	201	187	144	

Draw an ordered stem and leaf diagram to show this information.

4 Izzy wants to buy some music online. These are the prices of the albums on her wishlist.

£3.33	£4.21	£5.16	£2.77	£4.79	£5.71
£3.54	£5.11	£5.07	£4.08	£3.88	£6.40

a Draw an ordered stem and leaf diagram to show this information.

b Work out the median.

c Work out the range.

5 Visitor figures to a museum are categorised as adult visitors or child visitors. The table shows the daily visitors for the last two weeks.

Adult	42	76	83	55	87	64	33	51	60	43	60	12
Child	12	6	8	142	20	13	16	10	22	9	13	19

a Display this information in a back-to-back stem and leaf diagram.

b Which value could be deemed an outlier? Give a reason for this figure.

REVISION CHECKLIST

- The median is the middle value when values are put in order of size. For an even number of values, the median is the mean of the middle two values.

- The mode is the value that occurs most often.

- $\text{mean} = \dfrac{\text{sum of a set of values}}{\text{the number of values in the set}}$

- range = highest value – lowest value

- For grouped data:
  - Estimate the mean ($\bar{x}$) using $\bar{x} = \dfrac{\sum fx}{\sum f}$, where f represents the frequency and x represents the midpoint of the class interval.
  - The modal class is the class interval with the highest frequency.

- Stem and leaf diagrams can be used to record and display information. Remember to order the leaves and include a key.

- Back-to-back stem and leaf diagrams can be used to compare two sets of data.

- Data sets can be compared by working out the average and range of each data set.

Exam-style questions

1 Sachin plays for a hockey team. These are the numbers of goals scored by the team in the last ten games.

  0   2   0   3   1   0   4   8   3   2

 a Work out the median number of goals scored.

 b Work out the range of the number of goals scored.

 c Work out the mean number of goals scored.

2 Fiona is researching a new cat food. She tests the food with eight cats. Fiona measures and records the mass of each cat daily. These are their masses, in kilograms, on Monday.

  6   5.5   7   4.9   6.8   8.8   9.5   10

Work out the mean mass.

3 The stem and leaf diagram shows the amount of time, in minutes, that Hilary spent on telephone calls to clients.

4	0 1 1 2 8
5	2 3 4 5
6	0 8 8 8 9
7	0 1

Key 4 | 0 = 40 minutes

What was the modal call length?

4 Sandi buys 11 plants as gifts for her co-workers. Their heights, in centimetres, are:

31, 54, 33, 18, 51, 33, 48, 47, 56, 60, 37

 a Write down the modal height.

 b Work out the range.

 c Find the median height.

5 Connor makes and sells necklaces. He uses chains cut to different lengths for the necklaces. The table shows the lengths, to the nearest centimetre, that he has to use.

Length of chain (cm)	Number available
16	5
18	8
20	1
22	14
24	7

 a How many chains does Connor have?

 b Work out the mean length of his chains.

6 James buys a selection of different smoothies to try. The list below shows the sugar content of each smoothie, in grams.

41, 32, 46, 50, 40, 53, 56, 48, 46, 44

  a Work out the range.

  b Work out the median sugar content.

  c Work out the mean sugar content.

7 Lucia is answering this question:

Find the median of these numbers.

4   9   12   7   6   4   5

She gives the answer as '7' because she knows that the median is the middle number.

Is she correct? Explain your answer.

8 The label on a box of chocolate chips states the average number of chips in each box is 500 chips.

Tia checks this statement by counting the number of chips in ten boxes and working out an average.

The number of chips on each of the ten boxes is:

497   501   488   504   498   530   487   488   497   510

Which average supports the statement on the label? You must show your working.

9 A small business has eight employees.

The table shows their monthly salaries.

Job	Number of employees	Monthly salary, £
Cleaner	1	900
Sales assistant	4	1600
Supervisor	2	1950
Manager	1	5250

  a What is the modal monthly salary?

  b Work out the median monthly salary?

  c Work out the mean monthly salary.

10 There are 24 people in a care home. Their ages are:

98	85	87	75	100	69	76	82
93	94	78	77	91	90	79	83
77	69	84	78	88	76	74	69

  a Draw a stem and leaf diagram to show this information.

  b Work out the range of ages.

  c Work out the median age.

11 Shahla collected data on the vehicles that passed her house. The table shows her findings.

	Car	Van	Lorry	Bus	Total
Red	2	3	5	7	
Blue	15	3	9	2	
White	9	4	0	1	
Green	2	0	1	0	
Total					

a Copy and complete the table.

b What is the modal colour of vehicle?

12 Arty is the supervisor at work. He notices that Josh and Ruth are always late and starts to record how late they are each day. The data he collects for Josh is shown in the table.

Minutes late	Tally	Frequency
10	ℍℍ	4
11	ℍℍ ℍℍ ǀ	11
12	ǁ	2
13	ℍℍ ǁǁǁǁ	9

a What is the range in time for Josh?

b Find the modal time for Josh.

Arty finds the mode for Ruth to be 10 minutes and the range to be 7.

c Compare the range in time and the modal time for Josh and Ruth.

13 Rafiq writes for the local newspaper. He writes an article about homelessness. Rafiq records the number of words in each sentence.

Number of words in a sentence	Frequency
12	18
13	21
14	7
15	9
16	16
17	18
18	10

a Work out the range.

b Work out the mean number of words per sentence. Give your answer correct to two decimal places.

c Rafiq looks at an article on firework safety. The mean number of words per sentence was 16.4 and the range was 10.

Compare the mean and the range for the two articles.

14 Jess and Tess swim for a local swimming team. These are the times, in seconds, recorded for their latest training sessions.

Jess	78	79	77	84	77	78	78	80	81	79	79
Tess	75	91	82	78	80	81	84	76	78	79	76

a Draw a back-to-back stem and leaf diagram to show the race times.

b Find the median time for each girl and compare their performances.

15 There is a number on each of a set of five cards. This information relates to the five numbers on the cards. The cards are in numerical order.

The range is 8.

The median is 12.

The highest number is 15.

Write down one possible set of five numbers.

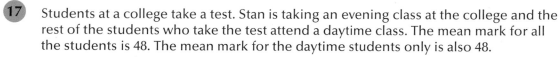

16 Eight numbers have a mean of 3.25.

What must the ninth number be so that the mean of the nine numbers is 3?

17 Students at a college take a test. Stan is taking an evening class at the college and the rest of the students who take the test attend a daytime class. The mean mark for all the students is 48. The mean mark for the daytime students only is also 48.

What does this tell you about Stan's score? You must explain your answer.

18 Give one advantage of using the mean as an average and one advantage of using the median as an average.

19 Find the average of the expressions on these cards. Give your answer in terms of x.

x $3x$ $3x + 5$ $4x - 2$ $5x - 7$ $2x + 8$

20 The table shows the marks scored by students in a test.

Mark, m	Number of students
$50 < m \le 60$	7
$60 < m \le 70$	12
$70 < m \le 80$	56
$80 < m \le 90$	64
$90 < m \le 100$	32

a How many students took the test?

b Work out an estimate for the mean mark.

**Now go back to the list of objectives at the start of this chapter.
How confident do you now feel about each of them?**

27 Time series and scatter graphs

Objectives

Before you start this chapter, mark how confident you feel about each of the statements below:

	▶	▶▶	▶▶▶
I can draw a scatter graph and interpret correlation as a relationship.			
I can draw a line of best fit and use it to interpolate and extrapolate apparent trends, recognising the dangers of doing so.			
I can identify outliers and explain why they may occur.			
I can construct line graphs for time series data and understand their appropriate use.			

Check-in questions

- Complete these questions to assess how much you remember about each topic. Then mark your work using the answers at the back of the book.

- If you score well on all sections, you can go straight to the Revision checklist and Exam-style questions at the end of the chapter. If you don't score well, go to the chapter section indicated and work through the examples and practice questions there.

1 The scatter graph shows ages and values of some cars. **Go to 27.1**

   a Draw a line of best fit on a copy of the diagram.

   b Estimate the age of a car when its value is £5000.

   c Estimate the value of a $3\frac{1}{2}$ year-old car.

   d Can the graph be used to predict the value of a 10-year-old car?

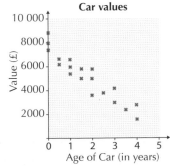

Car values

2 The table shows bike sales at a local store. **Go to 27.2**

Month	Number of bikes
March	3
April	6
May	10
June	12
July	14
August	10
September	5

a Draw a time-series graph for this data on axes similar to those shown.

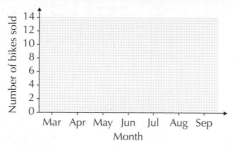

b Describe the trend in sales shown.

27.1 Scatter graphs and correlation

A **Scatter graphs** shows the relationship between two sets of data. The relationship is called correlation.

It is important to remember that **correlation** does not imply **causation**. For example, the number of firefighters at a fire usually increases as the fire spreads. Although there is a positive correlation, there is not causation; bringing more firefighters to fight the fire does not cause the fire to increase its spread.

Correlation

Positive correlation occurs when both variables are increasing. For example, the taller you are, the heavier you are likely to be. If the points are nearly in a straight diagonal line, as in the example below, there is a strong correlation.

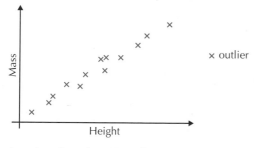

A point that does not fit in with the pattern of the data is known as an **outlier** and can be ignored if it is thought to be due to an error in measuring.

Negative correlation occurs when one variable increases as the other decreases. For example, as the temperature increases, the sales of woollen hats are likely to decrease.

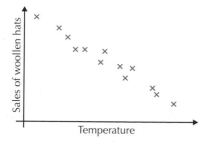

Zero correlation occurs when there is no linear relationship between the variables. For example, there is no connection between the height of a person and their mathematical ability.

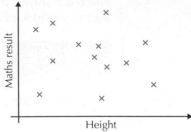

You can draw a **line of best fit** on a scatter graph that shows positive or negative correlation. It is drawn as close to all the points as possible with approximately half the points above the line and half the points below it.

You can use a line of best fit to esimate the value of an unknown variable from a known variable.

Predicting a value that is within the range of the data is called **interpolation.** This can be quite accurate.

Predicting a value that falls outside the range of the data is called **extrapolation.** These predictions are generally less accurate.

Example 1	**Q** The scatter graph below shows the percentage scores of some students in Science and Maths. A line of best fit has been drawn.

a What type of correlation is shown in the graph?

b A student was absent for the Maths test, but scored 30 on the Science test. Estimate her score.

c A student was absent for the Science test, but scored 50 on the Maths test. Estimate his score.

A **a** positive correlation   **b** 18   **c** 54

Exam tips   Check your plotted points carefully when you draw scatter graphs.
Draw a line of best fit when you are asked to estimate a value from a scatter graph. Note that a line of best fit does not need to pass through the point (0, 0).

Practice questions

1   Match each graph with the type of correlation that it shows.

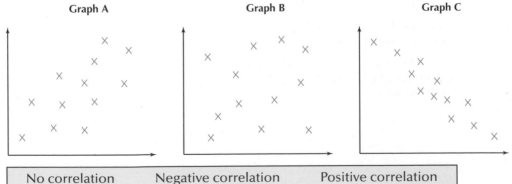

Graph A   Graph B   Graph C

No correlation   Negative correlation   Positive correlation

2 The table shows the number of pages and prices of books sold by a bookshop between 9 a.m. and 10 a.m. one Monday.

Number of pages	416	366	64	432	272	60	598	324	502
Price (£)	5.99	8.16	2.99	3.85	5.51	0.56	7.54	4.14	6.94

a Draw a scatter graph to show the data using 2 cm to represent 100 pages on the horizontal axis and 1 cm to represent £1 on the vertical axis.

b Draw a line of best fit.

c Estimate:
  i the cost of a book with 260 pages
  ii the number of pages in a book costing £1.50.

3 Miss Directed is investigating this claim about a particular manufacturer's motor cars.

'The older the car, the less it is worth.'

Miss Directed collects some data from a local newspaper and draws a scatter graph to show her results.

Miss Directed sees three more cars to add to her data:

3 years old, valued at £10 000

3 years old, valued at £9000

7 years old, valued at £5500

a Plot these points on a copy of the scatter graph.

b Does the data support the claim?

c Cars A and B do not fit the general trend. What can you say about:
  i car A
  ii car B?

d Do you think that this general trend continues as cars get even older? Explain your answer.

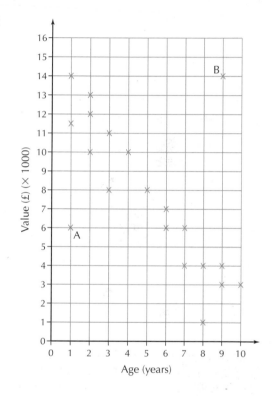

4 Mr Stephens is investigating this claim.

'The greater the percentage attendance at my Maths lessons, the higher the mark in the end of year maths test.'

Mr Stephens collects data from his class and draws a scatter graph to show his results.

a Decide whether or not the data supports the claim.

b Student A does not fit the general trend. What can you say about student A?

c Describe the correlation between percentage attendance and the maths test result.

d Draw a line of best fit on the scatter graph.

e Use your line of best fit to estimate:

   i the test result of a student who attended 85% of Maths lessons

   ii the attendance of a student with a test result of 60%.

f Why is estimating the test mark for a student who attended 40% of lessons unlikely to be accurate?

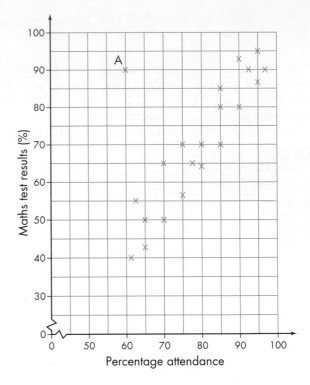

5 The table shows the scores of some students in two different tests.

Student	Amy	Billy	Chris	Dom	Eddy	Flynn	Glyn	Harjit	Iris	Jill
Mental	13	17	22	25	20	14	14	18	12	21
Written	25	30	34	13	37	29	25	44	19	35

a Plot the data on a copy of these axes.

b Draw a line of best fit.

c Keith missed the mental test, but scored 32 on the written test. Estimate his likely score on the mental sets of test results.

d One person did not do as well as expected in the written test. Who do you think that was? Give a reason for your choice.

e Describe the correlation between the two sets of test results.

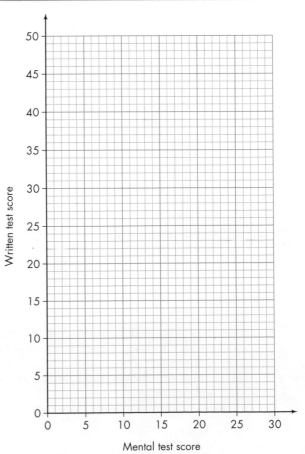

27.2 Time-series graphs

A **time-series** graph shows visually how a variable changes over time. This is sometimes called the trend of the data. They are used in hospitals for recording body temperature over a period of time, and in economics to show how the Retail Price Index, share prices, etc. change over time. Time is always plotted on the horizontal axis.

See Chapter 11 for distance–time graphs.

Example 2

Q Mark is in hospital. His temperature is measured every two hours and recorded in the table.

a Show this information on a time-series graph.

b Can you use this time-series graph to find Mark's exact temperature at 15:00? Explain your answer.

Time	Temperature (°C)
06:00	38.6
08:00	38.2
10:00	38.8
12:00	38.8
14:00	38.0
16:00	37.6
18:00	37.6
20:00	37.2

A

a

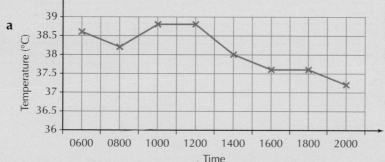

b You cannot use the graph to find the exact temperature at 15:00. The values in between the plotted points are not recorded values.

Exam tips Remember the values between plotted points are not recorded values. The points are joined to show the trend. In this type of question, there are usually marks available for communicating your answer.

Practice questions

1 Fill a glass with iced water, then measure and record the temperature every 5 minutes for half an hour. Construct a time-series graph to show your data.

2 The graph shows the temperature in my house, taken at different times of the day and night.

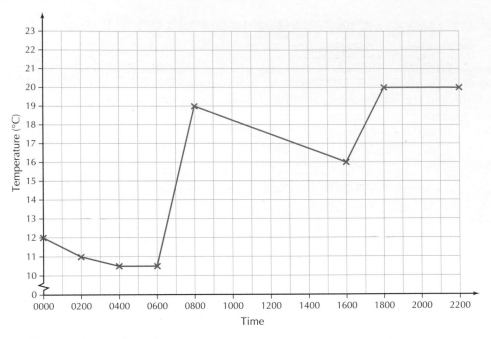

a Copy and complete the sentences.

At 12 noon, an estimate for the temperature in the house is ___° C. The house cooled down at the fastest rate between ___ and ___.

b Explain why you cannot say what the exact temperature was at midday.

3 Neil recorded his weight at the end of each two-week period for the first ten weeks of his diet.

Week	0	2	4	6	8	10
Mass (kg)	84.7	83.7	82.5	81.7	81.0	80.3

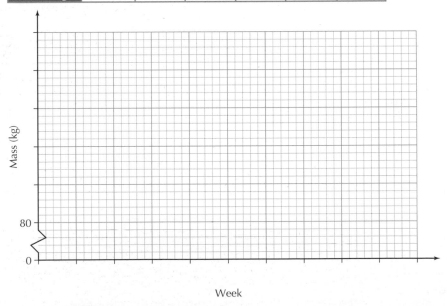

a Draw a line graph for the data.

b Use your graph to estimate Neil's mass at the end of the third week.

c In which two-week period did Neil lose most weight?

d Comment of the trend of the data.

e Can you predict Neil's mass at the end of week 14? Explain your answer.

4 The table shows the number of overseas visitors to London from 2012 to 2017.

Year	2012	2013	2014	2015	2016	2017
Number of visitors (millions)	14.8	14.2	14.7	15.3	15.5	16.8

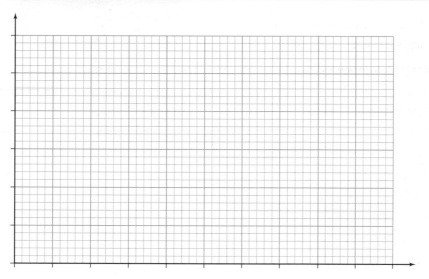

a Draw a line graph for the data.

b Estimate the number of visitors in 2018. Explain your answer.

c Between which two years did visitor numbers increase the most?

d Explain the trend in the number of visitors to London. What reasons can you give to explain this trend?

e Is it possible to use this data to predict the likely number of visitors in 2025?

REVISION CHECKLIST

- Positive correlation: both variables are increasing.
- Negative correlation: as one variable increases, the other decreases.
- Zero correlation: little or no linear correlation between variables.
- On a scatter graph, a line of best fit should be as close as possible to all the points in the direction of the data.
- Predicting a value that is within the range of the data is called **interpolation.** This can be quite accurate.
- Predicting a value that falls outside the range of the data is called **extrapolation.** These predictions are generally less accurate.
- Time-series line graphs show trends in the data. The points are plotted and joined with straight lines. Some form of time measurement is on the horizontal axis.

Exam-style questions

1 The table shows the price (to the nearest £) and age (to the nearest year) of some hardback books.

Age, years	0	1	2	10	16	25	50	75	90	100	110	120
Price, £	10	10	7	3	250	1	35	85	120	150	155	200

a Draw a scatter graph to show the data.

b Which value do you think is an outlier? Give a reason for your answer.

2 The table shows the times taken and the distances travelled by a delivery driver for eight deliveries.

Distance, miles	1.5	7.9	4.5	6.0	4.9	6.6	7.9	4.3
Time taken, minutes	4	15	8	12	7	15	16	5

a Draw a scatter graph to show the data.

b Describe the correlation.

3 The scatter graph shows how much money was donated by visitors to a museum over several days.

a What type of correlation is shown here?

b Explain what the graph shows.

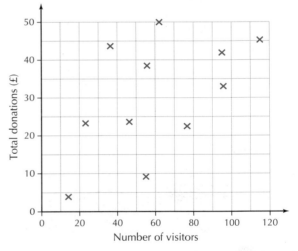

4 The scatter graph shows the lengths and masses of the fish Karen caught one Saturday morning.

a What is the likely mass of a fish that is 30 cm in length?

b What is the likely length of a fish with a mass of 4.5 kg?

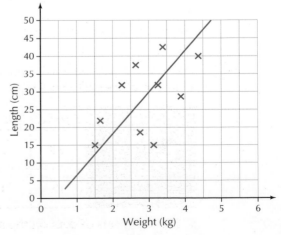

5 The scatter graph shows the costs and distances of some taxi journeys.

  a How much should you expect to pay for a journey of 4 km?

  b A taxi journey cost £6.50. What is the likely distance of the journey?

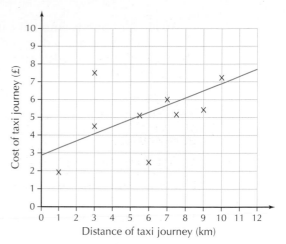

6 The graph shows the distance an aeroplane has travelled at various times on its journey. How many miles has the aircraft travelled after 2 hours?

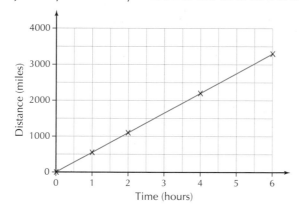

7 The normal depth of a swimming pool is 1 metre. The table shows the depth of water in an outdoor swimming pool at the end of each day for the first 10 days in June.

June	1st	2nd	3rd	4th	5th	6th	7th	8th	9th	10th
Depth of water (metres)	0.8	0.8	1	1	1.4	1.4	1.4	0	0.4	0.7

Complete each sentence with the correct date.

The pool was emptied on

It rained on

The pool was at the normal depth on

| 5th June |
| 3rd June |
| 8th June |

8 The table shows the maximum and minimum temperatures (to the nearest degree) in a conservatory each day for a week.

Day	Sunday	Monday	Tuesday	Wednesday	Thursday	Friday	Saturday
Maximum temperature, °C	22	24	29	25	15	14	14
Minimum temperature, °C	12	13	10	11	12	12	10

  a Plot both sets of data on the same axes.

  b Describe any similarities between the trends.

c An air vent opens automatically when the temperature in the conservatory reaches 25° C. On how many days did the vent open?

d If the temperature drops below 8° C, a heater comes on. On how many days did the heater come on?

9 The values and ages of some cars for sale at Orchard Garage are shown in the table.

Age, years	1	2	5	3	1.5	4	5	9
Price, £	8200	7500	5000	7500	9000	6000	6000	1000

a Copy and complete the scatter graph.

b Draw a line of best fit on your diagram.

c Pam wishes to sell her 6-year-old car. How much should she expect Orchard Garage to sell it for?

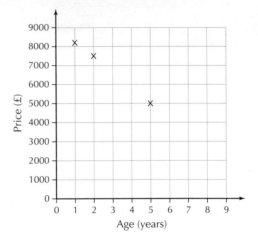

10 The table shows the prices charged by electricians for jobs taking different lengths of time.

Time taken, hours	1	3	1.5	2	4	6	4	2.5
Price, £	50	80	60	60	100	150	120	75

a Copy and complete the scatter graph.

b Draw a line of best fit on your diagram.

c Millie needs some work done that is likely to take 5 hours. Use your diagram to work out how much she should expect to pay.

d Libby needs a job done that will take 10 hours. Explain why you cannot use your diagram to estimate the cost.

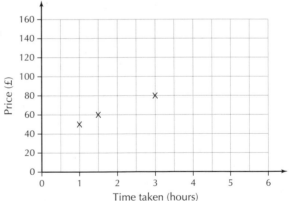

Now go back to the list of objectives at the start of this chapter.
How confident do you now feel about each of them?

Online stretch lessons

If you want to go further into some of the topics in this book, you can access lessons online at **www.collins.co.uk/edexcelpost16** covering the following objectives:

	▶	▶▶	▶▶▶
Chapter 7 stretch lesson: Equations and Inequalities			
I can solve linear equations involving fractions.			
I can solve quadratic equations by factorising.			
I can solve two inequalities and compare them to find values that satisfy both inequalities.			

Chapter 8 stretch lesson: Geometric sequences			
I can continue a geometric progression and find the term-to-term rule.			

Chapter 9 stretch lesson: Finding the equations of straight lines			
I can find the equation of a line through one point with a given gradient.			
I can find the equation of a line through two given points.			

Chapter 13 stretch lesson: Proportionality			
I can set up and use equations to solve problems involving direct and inverse proportion.			

Chapter 15 stretch lesson: Complex similarity problems			
I can solve complex similarity problems.			

Chapter 19 stretch lesson: Problem solving			
I can apply Pythagoras' theorem, bearings and trigonometry to more complex problems.			

Chapter 22 stretch lesson: Vectors			
I can calculate using column vectors and represent the sum and difference of two vectors graphically.			
I can identify column vectors that are parallel.			

Chapter 28 stretch lesson: Simultaneous linear equations			
I can write simultaneous equations to represent a situation.			
I can solve two linear simultaneous equations algebraically.			
I can solve two linear simultaneous equations graphically.			
I can solve simultaneous linear equations, representing a real-life situation and interpret the solution.			

Chapter 29 stretch lesson: Constructions			
I can construct the perpendicular bisector of a given line.			
I can construct a triangle.			
I can construct the perpendicular from a point to a line.			
I can construct the bisector of a given angle.			
I can construct angles of 90° and 45°.			
I can find and describe regions which satisfy a combination of loci.			
I can solve a variety of locus problems.			

Answers

Chapter 1

Check-in questions

1 a 33   36   363   3603   33 060

  b 120   521   1250   2501   12 005

  c −640   −406   46   64   4060   6004

  d −437   −73   407   3047   7340

2 a 5   **b** 512   **c** 7   **d** 2401

3 a $50 = 2 \times 5 \times 5$ or 2×5^2
  b $360 = 2 \times 2 \times 2 \times 3 \times 3 \times 5$ or $2^3 \times 3^2 \times 5$
  c $16 = 2 \times 2 \times 2 \times 2$ or 2^4

4 a false   **b** true   **c** true   **d** false

5 a 6^8 **b** 12^{13} **c** 5^6 **d** $4^2 = 16$

6 a 6.4×10^4 **b** 4.6×10^{-4}

7 a 1.2×10^{11} **b** 2×10^{-1}

8 a 1.4375×10^{18} **b** 5.48×10^{19}

1.1 Positive and negative integers

1 a $2 < 5$ **b** $2 > -5$ **c** $41 > 14$ **d** $-3 > -9$

2 a −5   **b** −9   **c** 7   **d** −7

3 c and **e**

4 −3° C

5 6 degrees

1.2 Square, cube and triangular numbers

1 b, **c**, **d** and **e**

2 a and **d**

3 a 9   **b** 4   **c** 1   **d** 10 000

4 21, 28

5 a $\dfrac{1}{5}$ **b** 3   **c** $\dfrac{9}{4}$
  d $\dfrac{20}{3}$ **e** $\dfrac{5}{26}$

1.3 Factors, multiples and primes

1 b, **c** and **d**

2 b, **d** and **e**

3 a $2^3 \times 3$ **b** $2^2 \times 3 \times 5$
  c $2^5 \times 3$ **d** $2^2 \times 5 \times 7$

4 a 12   **b** 4

5 a 840   **b** 480

6 2 p.m.

1.4 Indices

1 a 3^5 **b** 10^3 **c** 2^6

2 a 5^4 **b** 7^8 **c** 2^{20} **d** 3^6

3 a 6   **b** 8^5 **c** 7^2

4 9 or −9

5 a 1   **b** 8   **c** 7
  d 0.125   **e** 0.001

1.5 Standard index form

1 2×10^{-8}

2 a 10 000   **b** 120 000
  c 0.000 012 3   **d** 0.1234

3 3.3×10^{-25}

4 a 6.2×10^3 **b** 5.8×10^3 **c** 1.2×10^6 **d** 3×10^1

5 2×10^0 or 2 kg

Exam-style questions

1 258 or 285

2 5 − 10, −11 + 6 and −8 − −3

3 5^4

4 9 and 1

5 $-\dfrac{1}{343}$ or -0.0029 (2 s.f.)

6 $\dfrac{8}{25}$

7 No, it should be 8^7.

8 4

9 $2^2 \times 29$

10 60

11 5

12 1

13 $\sqrt{48}$

14 10×10^{-5}

15 2:10 a.m.

16 $a = 2$

17 6.08×10^5

18 300

19 18

20 $\dfrac{9}{4}$

Chapter 2

Check-in questions

1   $\dfrac{3}{8}$ $\dfrac{1}{2}$ $\dfrac{7}{12}$ $\dfrac{4}{5}$

2   $\dfrac{5}{9} \to \dfrac{20}{36}, \dfrac{3}{10} \to \dfrac{9}{30}, \dfrac{4}{5} \to \dfrac{24}{30}, \dfrac{1}{3} \to \dfrac{12}{36}$

3   **a** $\dfrac{13}{15}$ **b** $2\dfrac{11}{21}$ **c** $\dfrac{10}{63}$ **d** $\dfrac{81}{242}$

4   **a** $5\dfrac{7}{10}$ **b** $1\dfrac{53}{90}$ **c** $\dfrac{32}{75}$ **d**   14

5   28 pages

6   Total time = 3 hours 25 minutes

7   In order:

   $0.4, \dfrac{675}{1000}, \dfrac{3}{4}, \dfrac{8}{10}, 0.85$

2.1 Fractions

1   **a** $\dfrac{33}{10}$ **b** $\dfrac{22}{9}$ **c** $\dfrac{61}{11}$ **d** $\dfrac{313}{23}$

2   **a** $5\dfrac{1}{3}$ **b** $9\dfrac{3}{10}$ **c** $3\dfrac{3}{4}$ **d** $53\dfrac{1}{2}$

3   **a** $\dfrac{1}{6}$ **b** $\dfrac{2}{5}$ **c** $\dfrac{3}{8}$ **d** $\dfrac{1}{4}$

4   $\dfrac{48}{52} \to \dfrac{12}{13}, \dfrac{13}{39} \to \dfrac{1}{3}, \dfrac{21}{28} \to \dfrac{3}{4}$

5   $\dfrac{3}{11}$ $\dfrac{9}{16}$ $\dfrac{3}{5}$ $\dfrac{7}{10}$

2.2 Calculating with fractions

1   £58

2   **a** $\dfrac{31}{40}$ **b** $\dfrac{23}{28}$ **c** $2\dfrac{31}{40}$

3   **a** $\dfrac{1}{4}$ **b** $\dfrac{5}{66}$ **c** $\dfrac{4}{5}$

4   **a** $\dfrac{5}{12}$ **b** $\dfrac{1}{2}$ **c**   8

5   **a** $2\dfrac{4}{5}$ **b** $\dfrac{2}{7}$ **c** $1\dfrac{13}{80}$

2.3 Fraction problems

1   No, the total is $1\dfrac{1}{10}$.

2   No, the area of the garden is $2\dfrac{1}{4}$ m².

3   $\dfrac{3}{8}$

4   $\dfrac{2}{5}$

5   4 days

2.4 Fractions and decimals

1   **a**   0.8   **b**   0.45   **c**   0.1875   **d**   0.54

2   **a** $\dfrac{4}{5}$ **b** $\dfrac{18}{25}$ **c** $1\dfrac{7}{20}$ **d** $2\dfrac{219}{250}$

3   $\dfrac{1}{6}, \dfrac{4}{7}$

4   1.024   1.24   2.014   2.41   4.21

5   2.401   2.14   0   2.041   2.014

Exam-style questions

1   $\dfrac{1}{6}$

2   0.45

3   $\dfrac{9}{25}$

4   $\dfrac{3}{16}$

5   Jorges (= 0.85, Tim = 0.8125)

6   10

7   $-\dfrac{1}{45}$

8   $5\dfrac{17}{36}$

9   33

10   $\dfrac{7}{10}$ $\dfrac{5}{7}$ $\dfrac{4}{5}$ $\dfrac{29}{35}$

11   7272

12   $4\dfrac{7}{8}$

13   12.201   11.2   11.102   11.02   1.21

14   $\dfrac{29}{40}\left(\dfrac{7}{10} = \dfrac{28}{40}\right)$

15   $4\dfrac{29}{40}$

16   0.13 $\dfrac{3}{20}$ 0.2 $\dfrac{1}{4}$

17   $\dfrac{1}{7}$ $\dfrac{5}{6}$

18   $\dfrac{2}{15}$

19   $\dfrac{3}{8}$

20   £52.50 (6 tubs)

Chapter 3

Check-in questions

1

Start number	× 10	× 100	× 1000
75	750	7500	75 000
184	1840	18 400	184 000
3	30	300	3000
0.6	6	60	600
0.07	0.7	7	70
1.758	17.58	175.8	1758

2

Start number	÷ 10	÷ 100	÷ 1000
75	7.5	0.75	0.075
184	18.4	1.84	0.184
3	0.3	0.03	0.003
0.6	0.06	0.006	0.0006
0.07	0.007	0.0007	0.000 07
1.758	0.1758	0.017 58	0.001 758

3 a 579.11   **b** 647.51   **c** 2500.2   **d** 325

4 a 50   **b** 1   **c** 12   **d** 21

5 a 38.2   **b** 46.1111111
c 15.64090909   **d** 4.261904762

3.1 Multiplying and dividing by powers of 10

1 a 630, 63 000   **b** 42.1, 4210
c 3920, 392 000   **d** 0.65, 65

2 a 2.15, 0.215   **b** 0.63, 0.063
c 0.0097, 0.000 97   **d** 85.42, 8.542

3 $9.3 \times 0.001 = 0.0093$, $0.93 \div 10 = 0.093$, $93 \times 100 = 9300$, $93 \div 10 = 9.3$

4 $380 \div 100$

5 The answer will be bigger. This is because if you divide by a number smaller than 1, the answer is bigger than the starting value, e.g. $2 \div \frac{1}{2} = 4$ and $4 > 2$.

3.2 Written methods

1 a 5492   **b** 154.86   **c** 55.54   **d** 62.433

2 a 5014.8   **b** 451.08   **c** 12.8   **d** 5000

3 £8.16

4 13 certificates

5 £3.35

3.3 Order of operations

1 a 11   **b** 0   **c** 15   **d** 20

2 a 27   **b** 15   **c** 10   **d** 98

3 a $5 \times 6 + 2 = 32$ **b** $16 - (6 \div 2) = 13$
c $3 \times (15 - 1) = 42$

4 No, the correct answer is 14.24.

5 $6 + 4 \times 8 \rightarrow 38$, $52 - 3 \times 5 + 18 \rightarrow 55$, $5^2 + 4 \times 2 \rightarrow 33$, $5 + 4^2 \times 2 \rightarrow 37$

3.4 Using a calculator

1 a 81   **b** 216   **c** 6.5536   **d** 0.00032

2 a 0.2   **b** 1.5   **c** 0.55555555555
d 5   **e** 0.0526315…

3 1.89

4 $\frac{4.8 \times 12.6}{40 \div 25} \rightarrow 37.8$, $\frac{4.2 \times 5.6}{20 \div 2.5} \rightarrow 2.94$, $3 \times 9^2 - 119 \times 2 \rightarrow 5$

5 Neither of them, the correct answer is 54.

Exam-style questions

1 6200

2 $34\,000 \div 1000$

3 $42.1 \times 10 = 421$, $304 \times 0.1 = 30.4$, $56 \div 100 = 0.56$, $79 \div 0.01 = 7900$

4 $8000 \div 20 \rightarrow 400$, $20 \times 40 \rightarrow 800$, $200 \div 400 \rightarrow 0.5$, $40 \div 20 \div 20 \rightarrow 0.1$

5 $20 \times 20 \times 0.0001$ $48 \div 160$ $9.6 \div 3035 \times 0.01$

6 $432 + 96$ $(= 528, 521 - 15 = 506)$

7 £82.03

8 173 miles

9 £183.60

10 52 DVDs

11 32

12 10

13 $5 - 3 + 2 = 4$

14 False, $30 - 2 = 28$

15

	True	False
$3 - 4 \times 2 = -2$		✓
$5 + 6 \times 7 = 77$		✓
$3 + 2 \times 4 - 5 = 6$	✓	
$25 \div 5^2 = 1$	✓	

16 0.04

17 reciprocal of 4

18 $\frac{1}{52} < 2 \times 0.01$

19 She is wrong. For example, the reciprocal of $\frac{1}{2}$ is 2 and $2 > \frac{1}{2}$.

20 $5^{2.3}$ $3^{3.4}$ 4^3 2^7

Chapter 4

Check-in questions

1 3700

2 a true   **b** false   **c** true   **d** false

3 69.5 kg

4 22

4.1 Rounding

1

	Rounded to 1 d.p.	Rounded to 2 d.p.	Rounded to 3 d.p.
27.1647	27.2	27.16	27.165
893.4078	893.4	893.41	893.408
0.0071	0.0	0.01	0.007
−3.0414	−3.0	−3.04	−3.041

2

	Rounded to 1 s.f.	Rounded to 2 s.f.	Rounded to 4 s.f.
4096	4000	4100	4096
12 304	10 000	12 000	12 300
679	700	680	0679
134	100	130	0134

3 £91

4

Dog	Recorded mass, kg	Lower bound, kg	Upper bound, kg
Murphy	32	31.5	32.5
Kim	15	14.5	15.5
Toby	12	11.5	12.5
Dinky	26	25.5	26.5

5 16.1925 m²

4.2 Estimating

1 500   30   100   100   40

2 57.6, 64.9 and 60.4

3 4

4 1

5 a 50 mph   **b** £48

Exam-style questions

1

	Rounded to 1 d.p.	Rounded to 2 d.p.	Rounded to 3 d.p.
3.0492	3.0	3.05	3.049
54.0041	54.0	54.00	54.004
109.104	109.1	109.10	109.104
3.007789	3.0	3.01	3.008

2

	Rounded to nearest 10	Rounded to nearest 100	Rounded to nearest 1000
7341	7340	7300	7000
12 662	12 660	12 700	13 000
203 641	203 640	203 600	204 000
97	100	100	0

3 nearest 10

4 8000

5 1000

6 48.333 and 48.327

7 £432.45

8 Most values rounded down to the nearest pound, so each price and therefore the total was more than his estimate.

9

	Rounded to 1 s.f.	Rounded to 2 s.f.	Rounded to 3 s.f.
7341	7000	7300	7340
12 662	10 000	13 000	12 700
203 641	200 000	200 000	204 000
97	100	97	097

10 88.8, 90.4 and 94.95

11 a false   **b** false   **c** true   **d** false

12 11

13 172

14 12.5 kg, 13.5 kg

15 2.05 kg

16 £28

17 57.95 kg, 58.05 kg

18 22.75 cm²

19 Either boy with justification. Jack has between £11.50 and £12.49 and Zeke has between £12.35 and £12.44.

20 £268.92

21 $71.5 \leqslant V < 72.5$

22 $7.4 \leqslant d < 7.5$

Chapter 5

Check-in questions

1 **a** true   **b** false   **c** true   **d** false

2 **a** $4a$ **b** $8a + b$ **c** $8a - 7b$
   d $15xy$ **e** $4a^2 - 8b^2$ **f** $2xy + 2xy^2$

3 **a** $4bc - 2ab$ **b** d^4 **c** $15mn$

4 **a** $6b^{10}$ **b** $2b^{-16}$ **c** $9b^8$ **d** $\dfrac{1}{25x^4y^6}$

5 **a** -6.2 **b** 4.36 **c** 9

6 Casey is 42, her daughter is 11

5.1 Algebraic terms and conventions

1 $9a^2$

2 $a \times b \to ab,\ a \div b \to \dfrac{a}{b},\ a \times a \to a^2,\ a \times b \div 2 \to \dfrac{ab}{2},$
   $2 \times a \times b \to 2ab$

3 $y = 3x - 6$ and $y + 3x = -6$

4 **a** $3a$ **b** $a + b$ **c** $3a + 6b$ **d** $3ab$

5 $2 \times 7 \times m \to 14m,\ 14 \div m \to \dfrac{14}{m},\ 6 + 8 + m \to 14 + m,$
   $2 + 7 \times m \to 2 + 7m$

5.2 Simplifying expressions

1 **a** $10x$ **b** $3y - 2$ **c** $a^2 + 6a - 6$

2 **a** $30a^2$ **b** $63xy$ **c** $80b^3$

3 $5a + 3b - 2a - 6b \to 3a - 3b,\ 3a + 13b^2 - 16b^2 \to$
   $3a - 3b^2,\ 3b^2 \times 5a - 3a \to 15ab^2 - 3a,\ 3b^2 \times 2a \to 6ab^2$

4 $16y + 1$

5 $5a + 3b$

5.3 Laws of indices

1 **a** a^{10} **b** b^{11} **c** c^7 **d** $32d^5$

2 **a** a^4 **b** b^9 **c** c^{-1} **d** $2d^3$

3 **a** false   **b** false   **c** false   **d** true

4 $\dfrac{21a^3b^4}{7ab^2} \to 3a^2b^2,\ \dfrac{5a^2 \times b^2}{b} \to 5a^2b,$
   $\dfrac{16a^5 \times 2ab^3}{8a^4b^5} \to \dfrac{4a^2}{b^2},\ \dfrac{3ab^5 \times 4a}{6a^4b^2} \to \dfrac{2b^3}{a^2}$

5 $16x6y^{10}$

5.4 Substituting into formulae

1 **a** 8   **b** 20   **c** 65   **d** −10

2 **a** £5   **b** £8   **c** £9.50   **d** £14

3 **a** £4.50   **b** £5.50   **c** £6.10
   d £6.70   **e** £8

4 **a** £8.44   **b** £5.98   **c** £14.32   **d** £19.12

5 £430.06

5.5 Writing formulae

1 **a** $10x = 180$ **b** $x = 18°$ **c** $36°, 36°, 108°$

2 **a** $\dfrac{n}{2} + 4 = n + 1$ **b** $n = 6$

3 **a** $5x - 10 = 3x + 12$ **b** $x = 11$

4 9

5 6

Exam-style questions

1 $8 + 2x$

2 $6x + 6$

3 3

4 **a** formula   **b** expression
   c identity   **d** equation

5 $15xy + 3y^2$

6 $15y + 5$

7 Bev is correct, there are no like terms.

8 $11a + 1$

9 $x = 2$

10 $x = 2$

11 14

12 8

13 a^{15}

14 b^{-7}

15 c^1

16 2156

17 $4a^4b$

18 **a** $12x + 6 = 15x - 10$ **b** 14

19 **a** always true   **b** never true
   c always true   **d** always true

20 19.85 cm²

Chapter 6

Check-in questions

1 a $3t^2 - 4t$

  b $4(2x - 1) - 2(x - 4) = 8x - 4 - 2x + 8$
       $= 6x + 4$

2 a $x^2 + x - 6$ **b** $4x^2 - 12x$ **c** $x^2 - 6x + 9$

3 a $6x(2y - x)$ **b** $3ab(a + 2b)$
  c $(x + 2)(x + 2) = (x + 2)^2$ **d** $(x + 1)(x - 5)$
  e $(x + 10)(x - 10)$

4 a $y(y + 1)$ **b** $5pq(p - 2q)$
  c $(a + b)(a + b + 4)$ **d** $(x - 2)(x - 3)$

5 $u = \pm \sqrt{v^2 - 2as}$

6.1 Expanding single brackets

1 a $3x + 15$ **b** $7y - 14$ **c** $13z + 65$

2 a $-2x - 8$ **b** $4x + 14$
  c $-15x - 9y$ **d** $5x^2 + 20x$

3 a $5x + 11$ **b** $6y + 3$
  c $2m + 8$ **d** $-2m + 33$

4 He added the 3 and 4 instead of multiplying them and he forgot to multiply the –2 by 3.

5 $3(y + 2) - 2(y - 3) = 3y + 6 - 2y + 6 = y + 12$;
   $2(y + 6) - y = 2y + 12 - y = y + 12$

6.2 Expanding two pairs of brackets

1 a $y^2 + 5y + 6$ **b** $x^2 + 6x + 8$ **c** $m^2 + 8m + 15$

2 a $x^2 + 3x - 4$ **b** $y^2 + 4y - 12$
  c $m^2 + 6m - 16$ **d** $x^2 - 25$

3 a $3x^2 + 14x + 8$ **b** $25y^2 + 10y - 8$
  c $16y^2 - 62y + 21$

4 $(x - 2)(x + 2) \rightarrow x^2 - 4$, $(x + 3)(x - 1) \rightarrow x^2 + 2x - 3$,
   $(x + 4)(x - 2) \rightarrow x^2 + 2x - 8$, $(2x + 1)(x - 1) \rightarrow 2x^2 - x - 1$

5 She forgot to multiply the 2 and x and the –2 and x. She also added the 2 and –3 instead of multiplying them.

6.3 Factorisation

1 a $2(2x + 5)$ **b** $5(3x + 7)$
  c $2(7x - 60)$ **d** $6(x - 5)$

2 a $5(y + 2x + 5z)$ **b** $8x(2 - 6x + 7z)$
  c $7(2x - 6y + 9z)$
  d $3(-3m + 4n - 7p)$ or $-3(3m - 4n + 7p)$

3 a $(x + 2)(x + 3)$ **b** $(x + 8)(x + 2)$
  c $(x - 7)(x - 3)$ **d** $(x + 8)(x - 2)$

4 a $(x - 3)(x + 3)$ **b** $(y - 7)(y + 7)$
  c $(m - 1)(m + 1)$ **d** $(3t - 11y)(3t + 11y)$

5 She did not fully factorise the expression. x is also a factor and should be brought outside the brackets.

6.4 Rearranging formulae

1 a $a = x - 6$ **b** $a = y + 6$ **c** $a = 6 - z$

2 a $b = \dfrac{x - 6}{4}$ **b** $b = \dfrac{y + 3}{5}$ **c** $b = \dfrac{6 - z}{2}$

3 a $R = \dfrac{V}{I}$ **b** $b = \dfrac{2A}{h}$ **c** $g = \dfrac{P}{mh}$

4 $y = -12 - m$

Exam-style questions

1 **b** and **d**

2 $3r(p + 4q - 7r)$

3 $5(x - 40) = 5x - 200 \neq 5x - 40$

4 $6x + 14$

5 $2x^2 + 13x - 24$

6 $6y^2 + 14y - 12$

7 $16y + 20$

8 $m^2 - 4m + 4$

9 $m + 2$

10 $m = \dfrac{y - c}{x}$

11 $p = \dfrac{Wq}{r}$

12 $10x^2 + 7x - 12$

13 $3x + 12 + 2x - 6 = 5x + 6$,
   $5x + 5 + 1 = 5(x + 1) + 1$

14 $(x - 12)(x + 12)$

15 $(3p - 5q)(3p + 5q)$

16 $\dfrac{9C}{5} + 32 = F$

17 $4x^4 - 8xy + 8xy - 16y^2 = 4x^2 - 16y^2$

18 $r = \sqrt{\dfrac{S}{4\pi}}$

19 $r = \sqrt{\dfrac{3v}{4\pi}}$

20 $(x - 5)(x + 4)$

Chapter 7

Check-in questions

1 a $x = 8$ **b** $x = -5$

2 a $x = -3.25$ **b** $x = -0.5$

3 a $x = -0.5$ **b** $x = 17$ **c** $x = -0.2$

4 $x = 5.5$ cm, shortest side 6 cm

5 $-2, -1, 0, 1, 2, 3, 4$

7.1 Linear equations of the form $ax + b = c$

1 **a** $x = 15$ **b** $x = 7$ **c** $x = 49$ **d** $x = 54$

2 **a** $x = 8$ **b** $x = 5$ **c** $x = 6$ **d** $x = 8$

3 **a** $x = 49$ **b** $x = 24$ **c** $x = 10$ **d** $x = 10$

4 $3x = 24 \rightarrow x = 8$, $5x + 2 = 22 \rightarrow x = 4$,
$\frac{x}{2} + 18 = 38 \rightarrow x = 40$, $6x - 9 = -21 \rightarrow x = -2$

5 **a** $x = 3$ (for **b**, **c** and **d**, $x = 2.5$)

7.2 Linear equations of the form $ax + b = cx + d$

1 **a** $x = 1$ **b** $x = 3$ **c** $x = 3$

2 **a** $x = 4$ **b** $x = -1$ **c** $x = -1.5$

3 3

4 **a** $x = 1$ **b** $x = -2$ **c** $x = -1.5$

7.3 Linear equations with brackets

1 **a** $x = 5$ **b** $x = 3$ **c** $x = 3$

2 **a** $x = 11$ **b** $x = 9$ **c** $x = 0$ **d** $x = 2$

3 **a** $x = 1$ **b** $x = 0.2$ **c** $x = 2.5$ **d** $x = 1.5$

4 15 cm

5 Student's own response.

7.4 Equation problems

1 $x = \frac{44}{9}$

2 94°

3 $m = \frac{1}{3}$

4 $x = 4$

5 158°

6 $x = 3.4$

7.5 Inequalities

1 **a**

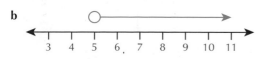

b

c

d

2 **a** $x < 3$ **b** $x > 19$
c $2 < x \le 5$ **d** $-8 \le x \le -1.5$

3 **a** $x > 6$ **b** $x < 20$ **c** $x \le 8$ **d** $x \ge 5$

Exam-style questions

1 $x = -5$

2 42°

3 $x = 4$

4 He should have added 4 to both sides.

$5x = 24 + 4$

$5x = 28$

$x = \frac{28}{5}$

$x = 5\frac{3}{5}$

5 $-4 < x \le 3$

6 **a** sides are equal **b** 40 units

7 $x = 2.5$

8 $x = 9$

9 $6x = 36$

10 $3x = 21 \rightarrow x = 7$, $x + 6 = 14 \rightarrow x = 8$,
$2x - 5 = 13 \rightarrow x = 9$, $4(x + 1) = 44 \rightarrow x = 10$

11

Chapter 8

Check-in questions

1 **a** 21, 34 **b** 36, 49 **c** 21, 28 **d** 13, 15

2 **a** $4n + 1$ **b** $2 - n$ **c** $2n$
d $3n + 2$ **e** $5n - 1$

3 **a** 2048, 8192 **b** multiply previous term by 4

4 $2n + 3$

8.1 Sequences

1 Odd numbers, because $2n$ is always even, so $2n + 1$ will be odd.

2 **a** 11, 18 **b** −6, −8 **c** 1.5, 2.5

3 1, 4, 9, 16... → square numbers ; 1, 3, 6, 10... → triangular numbers; 1, 8, 27, 64... → cube numbers; 1, 1, 2, 3, 5... → Fibonacci numbers

4 a 38   **b** 66
   c 80   **d** −25

5 8

8.2 Finding the *n*th term of an arithmetic sequence

1 4, 7, 10, 13

2 a $6n - 4$ **b** $4n - 5$
   c $8n + 2$ **d** $105 - 4n$

3 a 12, 24   **b** 21, 17   **c** 42, 84   **d** 5, 14

4 No, because if $2n^2 + 5 = 41$, $2n^2 = 36$ and $n^2 = 18$. Since 18 is not a square number, there is no integer value for n, so 41 is not a term in the sequence.

5 423

Exam-style questions

1 cube

2 11 and 19

3 32, 35

4 41, 38

5 prime numbers

6 add 4

7 194

8 5, 1, −3, −7

9 $6n + 3$

10 5, 40

11 a arithmetic or linear   **b** 74

12 $69 - 4n$

13 Yes, because $2n - 6 = 56$, $2n = 62$, $n=31$ which is a whole number.

14 $\dfrac{2n-1}{n+2}$

15 No, because for a common term $4n - 1 = 53 - 3n$ so $7n = 54$ and there is no integer value of *n* for which $7n = 54$.

16 a 10   **b** They are triangular numbers.

Chapter 9

Check-in questions

1 a

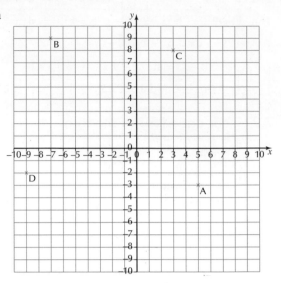

b AB = (−1, 3), AD = (−2, −2.5) and BD = (−8, 3.5)

2 a

x	−2	−1	0	1	2	3
y	−1	1	3	5	7	9

b

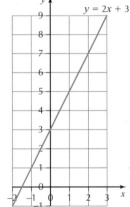

3 4

9.1 Coordinates

1 (3, 5)

2

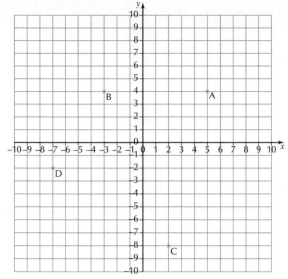

3

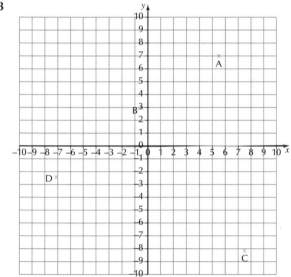

4 **a** (−0.85, 2.3) **b** (0.15, −5.25)
  c (−3.85, 0.3) **d** (3.5, −2.55)

5 B

9.2 Drawing straight-line graphs

1

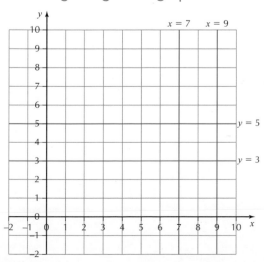

2 A: $y = 7$, B: $x = 2$, C: $y = 1$ and D: $x = 6$

3 a

x	−2	−1	0	1	2	3
y	1	2	3	4	5	6

b

x	−2	−1	0	1	2	3
y	−7	−5	−3	−1	1	3

c

x	−2	−1	0	1	2	3
y	3	2.5	2	1.5	1	0.5

4

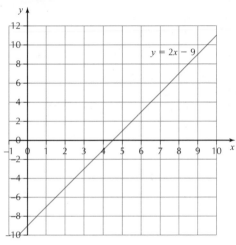

$x = 7$

5

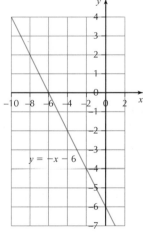

$x = −9$

9.3 Calculating the gradient of a straight-line graph

1 A = 0.5, B = 1, C = 2, D = 2, E = −2, F = −0.5

2 a 3 **b** 1 **c** ½ **d** −1

3 A + E, B + F, C + D

4 a 6 **b** −3 **c** −7 **d** 4

5 gradient = 2 and y-intercept = −3

Exam-style questions

1 $(5, 0)$

2

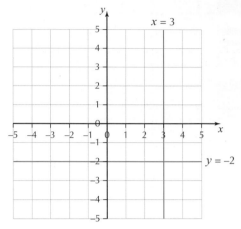

$(3, -2)$

3 **a** X   **b** W   **c** Z   **d** Y

4 Any line y = constant

5 $y = 5 \rightarrow$ horizontal, $x = -2 \rightarrow$ vertical,
$y = 3x - 4 \rightarrow$ positive gradient,
$2y = -2x - 1 \rightarrow$ negative gradient

6 $(4, 11)$

7 $(2.5, 8.5)$

8 $(4, 6)$

9 A, B, D and E

10 -2

11 D

12 gradient $= \dfrac{6}{5}$, y-intercept $= 2$

13 -2

14 $y = -0.5x + 7$

15 $x =$ any constant

16 A and C

17 $y = 2x - 5$

Chapter 10

Check-in questions

1 B: $y = 5x^2 - 9$

2

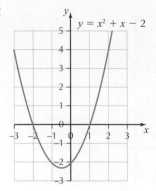

3 $1.85, -4.85$

4 Graph A: $y = \dfrac{3}{x}$

Graph B: $y = 4x + 2$

Graph C: $y = x^3 - 5$

Graph D: $y = 2 - x^2$

10.1 Quadratic graphs

1 $(0, -4)$

x	-2	-1	0	1	2	3	4
y	-1	-3	-3	-1	3	9	17

3 $x = -2$

4 a

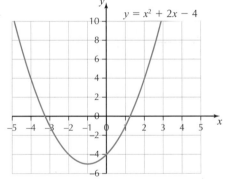

b

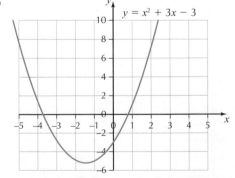

c

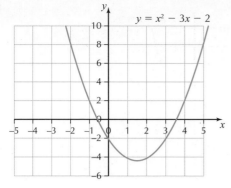

$y = x^2 - 3x - 2$

d

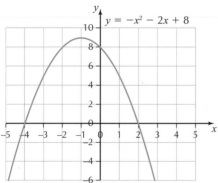

$y = -x^2 - 2x + 8$

5 a

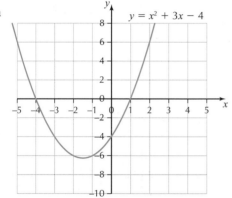

$y = x^2 + 3x - 4$

i $(-1.5, -6.25)$ **ii** $x = -1.5$

b

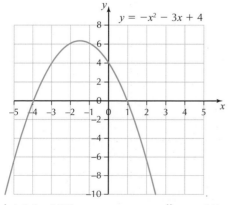

$y = -x^2 - 3x + 4$

i $(-1.5, -6.25)$ **ii** $x = -1.5$

c

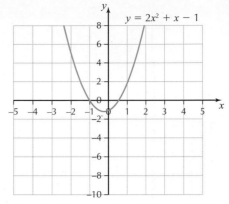

$y = 2x^2 + x - 1$

i $(-0.25, -1.125)$ **ii** $x = -0.25$

d

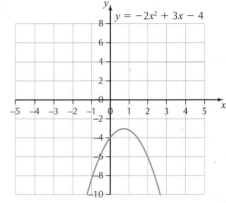

$y = -2x^2 + 3x - 4$

i $(0.75, -2.875)$ **ii** $x = 0.75$

10.2 Using graphs to solve quadratic equations

1 $x = -1$, $x = 3$

2 $x = 1$, $x = -2$

3 $x = 3$, $x = -4$

4 The solutions will be the coordinates of the point(s) of intersection.

5 The graph is above the x-axis, so does not cross it, hence it has no roots.

10.3 Graph shapes

1 Graph A $\rightarrow y = \dfrac{1}{x}$, Graph B $\rightarrow y = -x^3 + 1$, Graph C $\rightarrow y = 2x^2 + 1$, Graph D $\rightarrow y = \dfrac{1}{2x^2} + 2x + 1$

2 $y = 3x^2 + 1$, because the coefficient of x^2 is higher

3

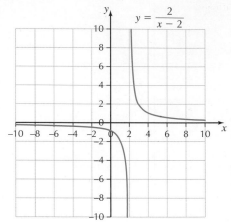

$y = \dfrac{2}{x - 2}$

b

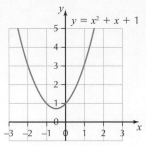

$y = x^2 + x + 1$

3

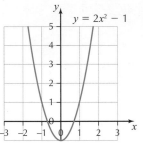

$y = 2x^2 - 1$

4 a

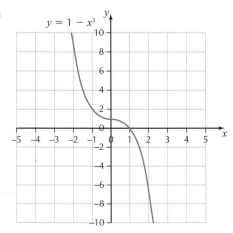

$y = 1 - x^3$

b $x \approx 1.8$

4

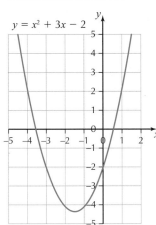

$y = x^2 + 3x - 2$

5 a

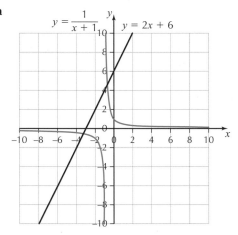

$y = \dfrac{1}{x + 1}$ $y = 2x + 6$

b $x = -3$, $y = -0.5$ and $x = -0.8$, $y = 4.5$

5

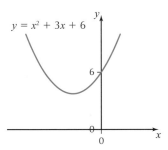

$y = x^2 + 3x + 6$

It's all above the x-axis.

6

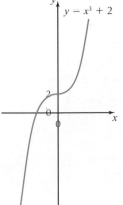

$y = x^3 + 2$

Exam-style questions

1

x	–2	–1	0	1	2
y	–6	–6	–4	0	6

2 a

x	–2	–1	0	1	2
y	3	1	1	3	7

1

7 0

8 $x = -1$

9 For example:

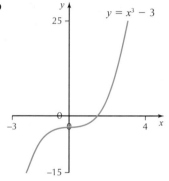

or

10 a

x	–2	–1	0	1	2	3
y	–11	–4	–3	–2	5	24

b

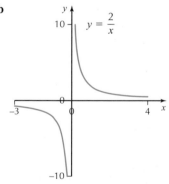

$y = x^3 - 3$

c 1

11 a

x	–3	–2	–1	–0.5	–0.4	–0.3	–0.2	0.2	0.3	0.4	0.5	1	2	3
y	$-\frac{2}{3}$	–1	–2	–4	–5	–6.7	–10	10	6.7	5	4	2	1	$\frac{2}{3}$

b

$y = \dfrac{2}{x}$

c not possible to divide by 0

12 A quadratic **B** reciprocal **C** cubic **D** none of these

13 A cubic **B** reciprocal **C** linear **D** none of these

14 For example:

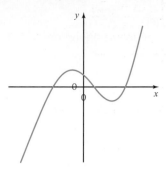

15 D

16 a $(x - 5)(x + 1)$ **b** $x = 5, x = -1$ **c** $(0, -5)$

d

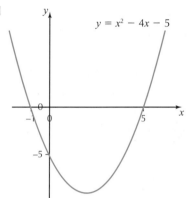

$y = x^2 - 4x - 5$

17 $x = -\dfrac{1}{4}$

18 a

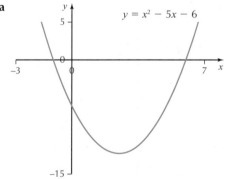

$y = x^2 - 5x - 6$

b $y = -9$

Chapter 11

Check-in questions

1 a 13.6 litres **b** 4.4 gallons

2 a 6 a.m. **b** 20° C

3 a 66.6̇ mph **b** 1 hour **c** 37.5 mph
  d 0820 **e** 50 mph

11.1 Conversion graphs

1 a £3.33   **b** $3.00
  c Find the number of dollars equivalent to £10, then divide by 10.
  d Find the number of dollars equivalent to £5, then multiply by 10.
  e In America, by $14 (or £9.33)

2 a

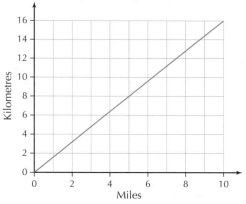

  b 8 km   **c** 7.5 miles
  d Student's own answer
  e Casey lives further away, because 38 km = 23.75 miles

11.2 Real-life graphs

1 a A = 4, C = 1, D = 5, E = 3, F = 2

  b B

2 No, the highest temperature was 20° C.

3 a 1 minute   **b** 3 minutes
  c 7 minutes   **d** 4 minutes

4 a £9.50   **b** £3.50   **c** 2 hours

5 The mouse population increased at an ever increasing rate.

11.3 Distance-time and velocity-time graphs

1 a i 12 noon   **ii** 3 p.m.
  b 160 miles
  c i 20 mph   **ii** 40 mph

2 a i 40 km/h   **ii** 5 km/h
      iii 0 km/h   **iv** 30 km/h

  b It has the steepest gradient.

3 a 6 km/h   **b** 0.6 m/minute
  c 0.8 km/h   **d** 1.4 km/minute

4 Rob runs at a constant speed throughout the race. Dan starts off at a faster speed than Rob, but then slows down. Just over half way through the

race, Rob overtakes Dan and although Dan speeds up again, Rob is first to finish.

5

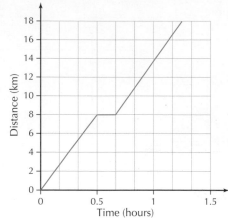

Exam-style questions

1 a 30 minutes   **b** 35 miles   **c** 26.25 mph

2 a 11 lb   **b** 4.5 kg

3 3 hours

4 Lucia. She has driven 21.9 miles and Toni has driven only 20 miles.

5 a 7 hours
  b No, the water level drops below 3 m before 5 a.m.

6 a 212° F
  b Jermaine is too hot, because 37° C = 98.6° F and his temperature is 102° F.

7 4 mph

8 a £8   **b i** 1   **ii** cost per km

9 £27.90

10 No, 110 kg = 17.3 stone

11 a At the start, for the first half a second, where the graph is the steepest.
  b 30 seconds   **c** 350 m/min

12 a Eddie   **b** 4 minutes   **c** Matthew

13 a

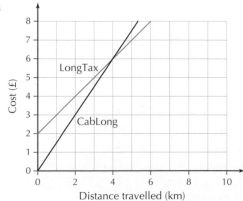

  b 4 km
  c LongTax, because it's cheaper by 50p.

14 a

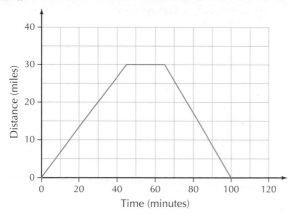

b 51.4 mph or 0.857 miles/min

15 a accelerating **b** constant speed

16 a 50 m/s **b** 3 s
  c The arrow slows down from its initial velocity of 50 m/s until it stops after 3 s.

Chapter 12

Check-in questions

1 a 6 kg **b** £600 **c** £3 **d** 252 g

2 a 0.4 kg **b** 5180 g **c** £4 980 000

3 £180

4 £42.25

5 80%

6 18.75%

7 51.7% (3 s.f.)

8 £121 856

9 £162.85

10 a £57.50 **b** £127 **c** £237.50 **d** £437.50

12.1 Percentage of a quantity

1 a 21 **b** 174 **c** 698 **d** 512

2 a 13 **b** 43.5 **c** 52.5 **d** 6156

3 a 175 **b** £418
  c 1127 kg **d** 214.32 m

4 a 46.2 g **b** 116.07 km
  c $10.42 **d** £183.60

5 a 60.27 m **b** 52.26 kg
  c £46.50 **d** 175.22 kg

12.2 Increasing and decreasing quantities by a percentage

1 a £169.50 **b** 420 g
  c 342.5 m **d** 1069.28 kg

2 a £32.76 **b** 11.04 kg
  c 236.39 km **d** 2468.48 mm

3 a 0.85 **b** 1.43 **c** 0.24 **d** 1.88

4 No, the cost of the jacket in the sale is £74.10.

5 Comp Solutions (total £534)

12.3 One quantity as a percentage of another

1 a 5% **b** 20% **c** 45% **d** 32.5%

2 a 20.0% **b** 42.9% **c** 27.1% **d** 32.8%

3 She did better in the first exam, 82.7%. Her score for the second exam was 76.2%.

4 ReadALot rated the book higher, 93% compared 90%.

5 5%

12.4 Percentage change

1 a 20% **b** 40% **c** 12.5% **d** 66.7%

2 a 56.5% **b** 41.7% **c** 12.5% **d** 95.7%

3 225%

4 48%

5 140%

12.5 Repeated percentage change

1 a £546.36 **b** £1461.28
  c £9115.50 **d** £12 873.77

2 214.4 g

3 598 737

4 12 years

5 £155.40

12.6 Interest and tax

1 £3.70

2 £132

3 £22.60

4 £93 400

5 Mandeep, because she has more salary left on which to pay tax.

12.7 Reverse percentage problems

1 £180

2 £1200

3 £15 000

4 £130

5 £78.67

Exam-style questions

1 41% of 50 is greater (20.5 compared with 20.4)

2 15

3 19.2%

4 43.2 g

5 1170 burgers

6 47.9%

7 32%

8 19.2%

9 27%

10 Rik (25%, Matt 17%)

11 26%

12 27%

13 217%

14 **a** 4200 steps   **b** 9 days

15 **a** 504.99   **b** 1 year

16 £5514.66

17 £4240

18 65.2 million

19 Bailey, £150

20 200 g

Chapter 13

Check-in questions

1 76 : 100

2 £20 : £40 : £100

3 £35.28

4 Large

5 3 days

13.1 Simplifying ratios

1 **a** 2 : 3   **b** 3 : 5   **c** 16 : 7   **d** 2 : 9

2 **a** 1 : 3   **b** 1 : 4   **c** 1 : 2   **d** 1 : 0.15

3 **a** $\frac{1}{3}$: 1   **b** $\frac{1}{4}$: 1   **c** $\frac{1}{2}$: 1   **d** $6\frac{2}{3}$: 1

4 15 : 20

5 1 : 12

13.2 Sharing a quantity in a given ratio

1 **a** £16 : 32   **b** £20 : 28
   c £12 : 36   **d** £40 : 8

2 Louise £35, Will £15

3 30 (20 dolls, 50 animals)

4 1600 females

5 £100.53

13.3 Problem solving

1 £1.99

2

£	$
10	**12.50**
12	15
25.50	**31.88**
38.72	48.40

3 10 crackers

4 10 kg sack

5 399 g

13.4 Increasing and decreasing in a given ratio

1 3 cups mashed potato, 1.5 onions, 1.5 eggs, 1.5 cups plain flour

2 6 weeks

3 7

4 No, he has enough for only 42

5 8.75 days

Exam-style questions

1 3 : 7

2 1 : 5

3 1 : 1.875

4 0.64 : 1

5 Sharne £105, Stevie £45

6 14 tubs

7 400 miles

8 £47.40

9 **a** $\frac{5}{13}$ **b** 61.5%

10 144°

11 Lily, since $19 = £15.20

12 Ken, since $15.60 = £12

13 £250

14 5 jewels

15 Yes, 186.6 g is needed

16 100 doors

Chapter 14

Check-in questions

1 3.2 kg

2 03:30

3 32.5 km

4 3.4 hours or 3 hours 26 minutes

14.1 Units of measurement

1 **a** tonnes   **b** metres   **c** centimetres
   d millimetres   **e** millilitres or litres

2

1 cm	=	10 mm	=	0.01 m
86 cm	=	**860 mm**	=	**0.86 m**
1 litre	=	**1000 cm³**	=	**100 cl**
56 kg	=	**0.056 tonnes**	=	**56 000 g**
206 m	=	**20 600 cm**	=	**0.206 km**

3 187.96 cm

4

12-hour clock time	24-hour clock time
1:58 p.m.	**13:58**
5:43 p.m.	17:43
2:06 a.m.	**02:06**
11:30 a.m.	11:30
9:15 a.m.	**09:15**
4:49 p.m.	16:49

5 0920

14.2 Maps and diagrams

1 60 km

2 **d** Lincoln and Stoke on Trent

3 **a, b, c** Check students' work

4

Actual distance	Distance on map
10 miles	2 cm
70 miles	**14 cm**
180 miles	**36 cm**

14.3 Compound measures

1

	Mass (g)	Density (g/cm³)	Volume (cm³)
a	3.6	**0.8**	4.5
b	54	1.2	**45**
c	**189**	3.15	60

2

	Force	Pressure	Area
a	**400 N**	20 N/m²	20 m²
b	106 N	**0.5 N/cm²**	212 cm²
c	50 N	1.25 N/cm²	**40 cm²**

3 33 180 kg

4

Speed	Distance	Time
100 m/s	250 m	**2.5 s**
60 mph	**210 miles**	3.5 hours
900 mph	450 miles	$\frac{1}{2}$ hour

Exam-style questions

1 1.5 mm is too thin

2 0.001 m

3 0.56 litres

4 0920

5 01:40

6 78 mm, 3.5 m, 4589 cm, 650 m, 5.6 km

7 1.830 litres

8

Actual distance	Distance on map
80 miles	8 cm
60 miles	**6 cm**
1000 miles	**1 m**

9 **a** 75 km   **b** 18 cm

10 14 cm

11 **a** 30 km   **b** 35.5 cm

12 4.8 cm

13

Speed	Distance	Time
100 m/s	350 m	**3.5 s**
65 mph	**227.5 miles**	3.5 hours
450 mph	150 miles	20 minutes

14 **a** 8 km/h   **b** 3 hours 7.5 minutes

15 **a** Check values and graph
    b 100 miles

16 30 minutes

17 160 km

18 0.25 N/cm²

19 540 mph

20 116 160 kg

Chapter 15

Check-in questions

1 a 13.8 cm   **b** 4.5 cm

2 20.7 cm (3 s.f.)

3 triangles A and E, SSS

15.1 Similar shapes

1 No, the ratios between the corresponding sides are not the same.

2 a Multiply the corresponding length in the smaller triangle by 3.

   b The corresponding angles will be the same.

3 $x = 8.4$, $y = 7.5$, $z = 92°$

4 x is an angle and corresponding angles are the same in similar triangles. Cannot work out the value of y, because no lengths are given so the scale factor is unknown.

5 larger triangle: 11.6 and 50°, smaller triangle = 5.8, 5.8, 80°

15.2 Congruent shapes

1 A, B, D and F, H, I

2 a congruent, SAS   **b** not congruent
   c not congruent   **d** congruent, SAS
   e congruent, AAS
   f congruent, AAS (as 2 angles are equal the third pair of angles must also be equal)
   g not congruent
   h not congruent

3 AAS [angle A = angle D = 40°; AC = DC; angle ACB = angle DCE (vertically opposite angles are equal)]

4 a AAS [angle ECD = angle ACB (vertically opposite angles are equal); angle EDC = angle ABC (alternate angles are equal); ED = AB (given)]
   b CD = 7 cm, CE = 5 cm

5

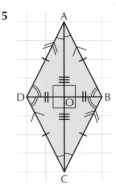

AOD, AOB, COB, COD are congruent triangles.

ADC and ABC are congruent triangles.

ABD and CBD are congruent triangles.

Exam-style questions

1 4

2 E

3 9.6

4 0.4

5 P: 15, Q: 8

6 a false   **b** false   **c** false   **d** false

7 a 15
   b No, the area will be $6 \times 15 = 90$.

8 a 5   **b** 6   **c** ii

9 a 10   **b** 42

10 Congruent, AAS [angle ACB = angle ECD (vertically opposite angles are equal); angle CAB = angle CED (alternate angles are equal), the corresponding sides between these angles are equal]

11 Congruent, AAS. Angle POQ = angle SOR (vertically opposite); angle QPO = angle SRO (given); corresponding sides PQ = SR (given).

12 a congruent, AAS [angle E = angle B (alternate angles are equal); angle EFD = angle BCA = 90° (given), angle A = 180 – (angle B + 90°) and angle D = 180 – (angle E + 90°) (angles in a triangle add up to 180°) so the triangles have equal angles. AB = DE]

   b AC = 6 cm

13 Yes, the corresponding sides are in the same ratio and angle C is common.

Chapter 16

Check-in questions

1 a a and b **b** a and c, b and c

2 hexagon

3 ABC: isosceles, one line of symmetry *or* one pair of equal sides/angles

   DEF: equilateral, 3 lines of symmetry *or* 3 equal sides/angles

4 a For example:

b All sides are different lengths, all angles are different sizes, no lines of symmetry.

5 a Rhombus **b** Kite

6 a and b

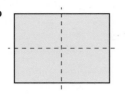

Order 2

c

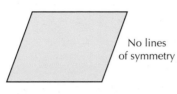

No lines of symmetry

Order 2

7 a i 12 **ii** 8 **iii** 6

b

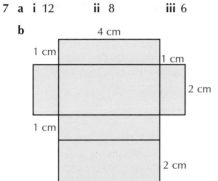

8 a

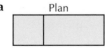

Plan

b Elevation

16.1 2D shapes

1

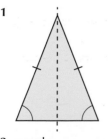

2 a and c

3 a AB, FC and ED; BF and CE, BC and EF; CD and AF

b Yes, CD = DE because they are two sides of a regular hexagon.

c Isosceles trapezium (as it is a trapezium with two equal sides)

4 a false **b** true **c** false
d true **e** true **f** false

5 Square, rectangle, rhombus

6 Square, rhombus, kite

16.2 Symmetry

1

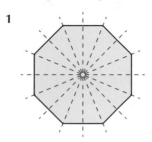

2 Yes

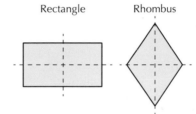

Rectangle Rhombus

3

4 Yes

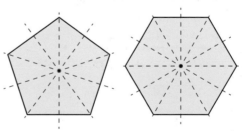

5 a **b** **c** **d**

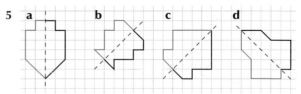

6 a i 0 **ii** 4 **b i** 4 **ii** 4
  c i 0 **ii** 6 **d i** 6 **ii** 6

16.3 3D shapes

1

Shape	Number of faces	Numbers of edges	Number of vertices
Cube	6	12	8
Triangular prism	5	9	6
Square-based pyramid	5	8	5
Tetrahedron	4	6	4

2 **a** and **c**

3

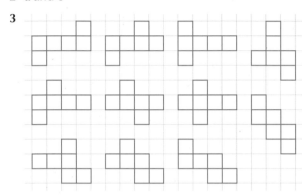

4

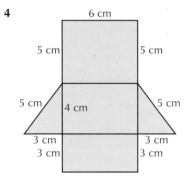

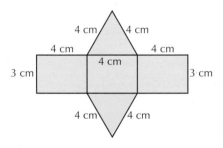

5 12 vertices, 8 faces, 18 edges

16.4 Plans and elevations

1 a

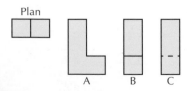

b

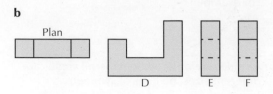

c

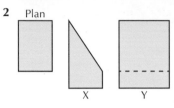

2

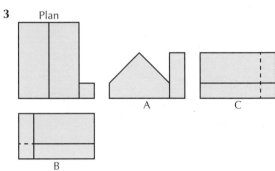

3

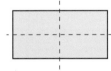

Exam-style questions

1 Equilateral triangle

2 a [BC, DA, EH, FG], [CD, HG], [DE, AH], [EF, BA]
b Isosceles   **c** Isosceles trapezium
d Rectangle

3 a   **b** order 2

4 a X   **b** V, W or Y
c Z   **d** X or Z

5

6

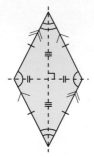

7 a

   b Rotational symmetry of order 2

8 a Regular decagon   **b** 10

9

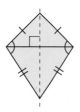

One line of symmetry.

Diagonals intersect at 90°.

Two pairs of equal sides. One pair of equal angles

10 For example

11

Rhombus	Parallelogram	Kite

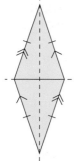

  Order 2   Order 2   Order 1 – no rotational symmetry

12 a C, F, and H   **b**

13

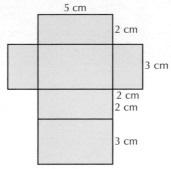

14 a Octagon

   b 16 vertices, 24 edges, 10 faces

15

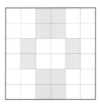

16 For example

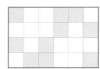

17 a i 8   **ii** 12   **iii** 6

   b For example

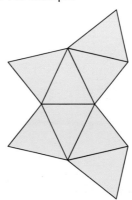

18 a 5 vertices, 9 edges, 6 faces

   b For example

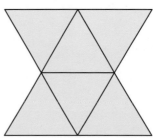

19 a

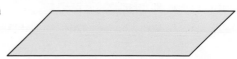

b

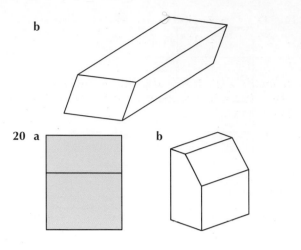

20 a

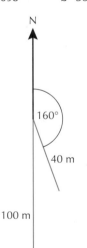

b

Chapter 17

Check-in questions

1 a right angle   **b** reflex   **c** reflex
   d acute   **e** obtuse

2 a 70°   **b** 150°   **c** 220°   **d** 290°

3 a $a = 55°$ **b** $a = 74°, b = 32°$

   c $a = 41°, b = 41°, c = 41°, d = 139°$

4 30°

5 a 072°   **b** 208°   **c** 290°   **d** 139°

6 a 098°   **b** 360° − 173° = 187°

7 a   **b** 192° for 64 m

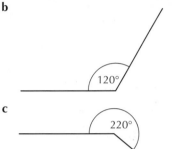

17.1 Classifying angles

1 a acute   **b** acute   **c** obtuse
   d obtuse   **e** obtuse

2 a 140°   **b** 40°   **c** 220°   **d** 321°

3 a acute   **b** right angle   **c** reflex
   d reflex   **e** obtuse

4 a 3°, 14°, 72°, 83°, 89°   **b** 172°, 91°, 100°
   c 240°, 320°, 350°, 230°, 190°, 300°

5 a

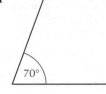

b

c

17.2 Angle facts

1 a $a = 157°$ (angles on a straight line add up to 180°)

   b $b = 60°$ (angles on a straight line add up to 180°)

   c $c = 130°$ (angles on a straight line add up to 180°), $d = 50°$ (vertically opposite angles are equal), $e = 130°$ (vertically opposite angles are equal or angles on a straight line sum to 180°)

   d $f = 150°$ (angles on a straight line add up to 180°), $g = 30°$ (vertically opposite angles are equal), $h = 70°$ (angles around a point add up to 360°)

2 a $a = 37°$ (angles in a triangle add up to 180°)

   b $b = 70°$ (angles in a triangle add up to 180°)

3 a $a = 30°$ (base angles of an isosceles triangle are equal), $b = 120°$ (angles in a triangle add up to 180°)

   b $c = d = 50°$ (base angles of an isosceles triangle are equal, angles in a triangle add up to 180°)

   c $e = 68°, f = 68°$ (base angles of an isosceles triangle are equal, angles in a triangle add up to 180°)

4 a $a = 80°$ (angles in a quadrilateral add up to 360°)

   b $b = 15°$ (angles in a quadrilateral add up to 360°)

   c $c = 70°$ (angles in a quadrilateral add up to 360° or angles in a triangle add up to 180°), $d = e = 20°$ (base angles of an isosceles triangle are equal, angles in a triangle add up to 180°)

5 a *a* =70° (base angles of an isosceles triangle are equal), *b* = 110° (angles on a straight line add up to 180°), *c* = 40° (angles in a triangle add up to 180°)

b *d* = 70° (vertically opposite angles are equal), *e* = 70° (base angles of an isosceles triangle are equal), *f* = 40° (angles in a triangle add up to 180°)

c *g* = 50° (vertically opposite angles are equal), *h* = 65° and *i* = 65° (angles in a triangle add up to 180°, the base angles of an isosceles triangle are equal)

17.3 Angles in parallel lines

1 Angles *a* and *c* are vertically opposite angles.

Angles *b* and *n* are corresponding angles.

Angles *b* and *h* are alternate angles.

Angles *b* and *f* are corresponding angles.

Angles *g* and *i* are alternate angles.

Angles *j* and *l* are vertically opposite angles.

Angles *p* and *d* are corresponding angles.

Angles *n* and *c* are allied angles.

2 a *a* = 130° (allied angles add up to 180°), *b* = 50° (corresponding angles are equal), *c* = 50° (alternate angles are equal)

b *x* = 60° (allied angles add up to 180°)

c *x* = 40° (corresponding angles are equal)

3 For example: Angle CBE = 40° (angles on a straight line add up to 180°), Angle FEH = 40° (corresponding angles are equal)

Angle BEF = 140° (corresponding angles are equal), Angle FEH = 40° (angles on a straight line add up to 180°)

Note: There are other methods that involve more steps.

4 Angle FEB =75° (corresponding angles are equal), Angle GED =75° (vertically opposite angles are equal)

5 Angle EHI = 55° (angles on a straight line add up to 180°)

Angle DEH = 55° (alternate angles are equal)

Angle BED = 45° (alternate angles are equal)

Angle BEH = 100° (Angle BEH = Angle BED + Angle DEH)

17.4 Angles in polygons

1 a The angles in a triangle add up to 180°. There are 3 triangles in the pentagon. 3 × 180° = 540°
b 540° ÷ 5 = 108°

2 a 115°   **b** 115°   **c** 182°

3 a 87°   **b** 55°   **c** 125°

4 8

5 If interior angle is 160°, the exterior angle is 180° – 160° = 20° and there are 360° ÷ 20° = 18 sides in the polygon.

6 70°

17.5 Bearings

1 SE or 135°

2 a 120°   **b** 065°   **c** 310°   **d** 230°

3 a 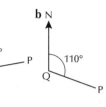 **b** 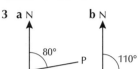 **c** **d**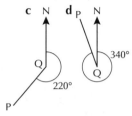

4 a 320°   **b** 260°   **c** 255°   **d** 120°

5 205°

17.6 Scale drawings and bearings

1 a Check student's answer.

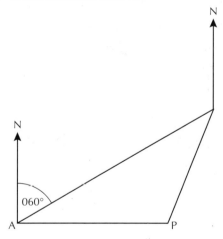

b bearing 252°, distance 19.4 km

2 a Check student's answer.

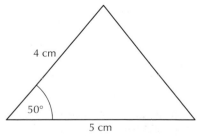

b 3.9 cm

3 21.8 km

4 a

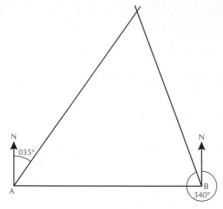

b 0.15 km closer

5 a 304°   **b**

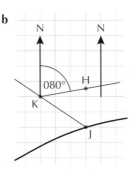

Exam-style questions

1 57.1 m

2 SW or 225°

3 122°

4 18°

5 32°

6 a 115° (angles in a triangle add up to 180°)
b 92.5° (angles in a straight line add up to 180°; angles in a quadrilateral add up to 360°)

7 $x = 45°$ (angles in a straight line add up to 180°), $y = 82°$ (angles in a triangle add up to 180°).

8 $x = 22$

9 50°

10 10°

11 He measured anticlockwise from North instead of clockwise.

12 2.7 cm

13 $x = 63$

14 No. If AB and DC were parallel, then the angles at B and C would be allied and would sum to 180°, but 110° + 60° = 170°.

15 Boat C is closer to Boat B by about 1.8 miles. (CB is 11.1 miles and AB is 12.9 miles)

16 Angle BFE = 50° (angles on a straight line add up to 180°)

Angle BEF = 80° (alternate angles are equal)

Angle EBF = 50° (angles in a triangle add up to 180°)

Angle BFE = angle EBF so triangle BEF is isosceles.

17 If EF is parallel to CD and AB, then angle ACE = 110° (68° + 42°, alternate angles are equal). Since angle ACE = 115°(given), EF cannot be parallel to CD and AB.

18 Angle ADC = 40° (allied angles add up to 180°), so $x = 15°$

Angle DCP = 180° − (25° +75°) = 80° (angles in a triangle add up to 180°)

Angle ADC + angle BCD = 180° (allied angles add up to 180°), so angle DCB = 140°

y = angle BCD − angle DCP

= 140° − 80°

= 60°

19 20

20 a 150°   **b** 30°   **c** 12

Chapter 18

Check-in questions

1 b

2 Yes, because $13^2 = 12^2 + 5^2$ (169 = 144 + 25)

3 a 8.1 cm   **b** 7.4 cm

4 $\sqrt{61}$

5 9.54 m

18.1 Calculating an unknown length

1 a 13   **b** 20   **c** 41

2 a 6.6   **b** 7.1   **c** 9.9

3 Yes, because it satisfies Pythagoras' theorem ($25^2 = 24^2 + 7^2$, 625 = 576 + 49)

4 a 13.1   **b** 10.4   **c** 48.5

5 a $\sqrt{41}$ **b** 8   **c** 7.5

18.2 Calculating the length of a line segment from the coordinates of its end points

1 6.4 units

2 7.8 units

3 a 4.24 units   **b** 8.60 units   **c** 11.3 units

4 8.54 units

5 a 7.62 units   **b** 8.49 units   **c** 7.07 units

18.3 Solving more difficult problems

1 6.93 units

2 £144

3 7.30 cm

4 3.32 m

5 3 m 12 cm

Exam-style questions

1 13.4 units

2 7.58

3 3.71 m

4 $y^2 = x^2 + z^2$, $x = \sqrt{y^2 - z^2}$ and $z = \sqrt{y^2 - x^2}$

5 50

6 28 m 28 cm

7 33.4 cm

8 20.9

9 75 cm

10 11.7 units

11 25.6 cm

12 13.4 cm

13 46.12 m

14 129.4 cm

15 6 cm

16 d, if the square has side 12 cm, then $d = \sqrt{288}$ and $h = \sqrt{192}$

17 $AB = \sqrt{17}$, $BC = \sqrt{17}$ and $AC = \sqrt{50}$. Two sides are equal so triangle ABC is isosceles.

18 4.47 units

19 Yes, $AB = \sqrt{32}$, $BC = \sqrt{18}$, $AC = \sqrt{50}$, so $AC^2 = AB^2 + BC^2$.

20 $WX = XY = YZ = ZW = \sqrt{17}$, so WXYZ must be either a rhombus or a square. $WY = XZ = \sqrt{34}$, so the diagonals are the same length. $WY^2 = WX^2 + XY^2$ (34 = 17 + 17) so the angle at X is a right angle. This is also true for the other vertices, so WXYZ is a square.

Chapter 19

Check-in questions

1 $\sin 60 = \dfrac{\sqrt{3}}{2}$ $\sin 45 = \dfrac{\sqrt{2}}{2}$

$\cos 30 = \dfrac{\sqrt{3}}{2}$ $\cos 0 = 1$

$\cos 90 = 0$ $\tan 45 = 1$ $\tan 60 = \sqrt{3}$

$\tan 30 = \dfrac{\sqrt{3}}{3}$

2 a 5.79 cm   **b** 8.40 cm

3 a 38.7°   **b** 43.0°

4 a 57.4°   **b** 26.5 m (3 s.f.)

5 12.4 km on a bearing of 218°

19.1 Trigonometric ratios

1 a 30°   **b** 60°

2 a 45°   **b** 30°

3 a 30°   **b** 60°

4 a 2   **b** $\sqrt{2}$

5 a 45°   **b** 30°   **c** 60°

19.2 Calculating a length

1 a $5\sqrt{3} = 8.66$ m   **b** 6 m   **c** $15\sqrt{3} = 26.0$ cm

2 a 0.574   **b** 0.174   **c** 0.466

3 a 7.66 m   **b** 14.5 cm   **c** 23.8 cm

4 a 2.5 cannot be the correct answer since x is the hypotenuse so should be the longest side.

     b The mistake is in the second line of working. It should read $x = \dfrac{5}{\sin 30}$.

5 a 8.5 m   **b** 8 cm   **c** 21.3 m

19.3 Calculating an angle

1 a 60°   **b** 30°   **c** 45°   **d** 30°

2 a 53.1°   **b** 18.7°   **c** 75.5°
   d 23.1°   **e** 24.2°   **f** 58.0°

3 a 44.4°   **b** 23.6°   **c** 50.2°
   d 61.9°   **e** 36.9°   **f** 36.9°

4 21.8°

5 28.56 m

6 14.8 cm

Exam-style questions

1 **d** tan 60 is greater than 1

2 30 × cos 40

3 35°

4 7.7 cm

5 2.95 m

6 29.7°

7 21.8°

8 224 m

9 26.8 m

10 bearing 163°, distance 20.9 miles

11 **a** 28.3 m **b** 058°

12 17.0

13 68.2°

Chapter 20

Check-in questions

1 **a** 12 cm, 6 cm² **b** 34 cm, 60 cm²
   c 26 cm, 36 cm² **d** 28 cm, 30 cm²

2 **a** 38 cm² **b** 35.705 cm²

3 **a** circumference = 31.42 cm,
    area = 78.55 cm² (2 d.p.)

   b circumference = 47.13 cm,
    area = 176.74 cm² (2 d.p.)

   c circumference = 43.99 cm,
    area = 153.96 cm² (2 d.p.)

4 perimeter = 41.13 cm, area = 100.53 cm² (2 d.p.)

5 **a** 14.05 cm **b** 49.17 cm²

20.1 Areas of plane shapes

1 **a** 32 m, 60 m² **b** 26 m, 32 m²
   c 23 cm, 24 cm²

2 **a** 24 m² **b** 30 m² **c** 48 cm²
   d 30 m² **e** 12 m²

3 C (A = 48 cm², B = 45 cm², C = 50 cm²)

4 8 cm

5 81 cm²

20.2 Areas of complex shapes

1 **a** 16 m² **b** 52 m² **c** 100 m²

2 14 m²

3 £33

4 £436

5 £476

20.3 Circles

1 **a** 31.4 cm **b** 11.3 mm
   c 62.8 cm **d** 22.6 mm

2 **a** 78.5 cm² **b** 10.2 mm²
   c 314.2 cm² **d** 40.7 mm²

3 257.10 cm²

4 **a** 21.46 cm² **b** 85.84 cm² **c** 23.18 cm²

5 **a** **i** 128π cm² **ii** 64π cm² **iii** 32π cm²
   b 16π cm²

20.4 Arc length and sector area

1 **a** 20.9 cm **b** 8.38 cm **c** 26.2 cm

2 **a** 6π cm² **b** 27π cm² **c** $\frac{24}{5}\pi$ m²

3 78.54 cm², 35.71 cm

4 **a** 88.36 m² **b** $\frac{45}{4}\pi + 10$ m

5 2.82 cm

6 25 laps

Exam-style questions

1 **a**   **b**   **c**

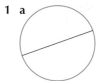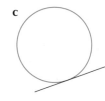

2 72 cm²

3 63 m²

4 2.5 m

5 20 cm²

6 50 m²

7 **a** £56
   b It may increase the cost, because if each tin (on average) covers less than 4.62 m², she will need to buy an extra tin.

8 h = 3 cm

9 1085.8 cm²

10 157.1 cm²

11 42.8 cm

12 401.7 cm²

13 17.7 cm

14 23.6 cm

15 31.7 cm²

16 47.1 cm

17 10.6 cm

18 92.7 cm²

19 15.96 cm

20 300π cm² *or* 942.5 cm²

Chapter 21

Check-in questions

1 a 96 cm³   **b** 388.8 cm³   **c** 1292.3 cm³

2 2.61 m

3 1 000 000

4 a 0.25 m²   **b** 2500 cm²

5 a 565 cm³   **b** 637 cm³   **c** 905 cm³   **d** 245 cm³

21.1 Volumes of prisms

1 150 cm³

2 261 cm³

3 192 cm³

4 1100 cm³

5 288 cm³

6 37 699 cm³

7 volume of A is $10 \times 3 \times 5 = 150$ cm³; volume of B is $0.5 \times 4 \times 5 \times 30 = 300$ cm³; 300cm³ = 2 × 150 cm³

21.2 Volumes and surface areas of more complex solids

1 $\frac{500}{3}\pi = 523.6$ cm³

2 300 m³

3 $45\pi = 141.4$ cm²

4 986 g

5 a $84\pi = 263.9$ cm³   **b** $54\pi = 169.6$ cm²

21.3 Converting units of area and volume

1 No, 0.5 m = 50 cm so the area is 1000 cm².

2 a 7200 cm²   **b** 0.72 m²

3 a 1000 cm³   **b** 0.001 m³

4 a 112 500 cm³   **b** 0.1125 m³   **c** 281.25 kg

Exam-style questions

1 252 cm³

2 1080 cm²

3 540 litres

4 14 cm

5 5 cm

6 1.09 g/cm³

7 1701 cm³

8 70 cm³

9 a 180 boxes   **b** It would increase to 200.

10 No, the volume of the bed is 660 000 cm³ $(110 \times 200 \times 30) = 660$ litres. This means she will need to buy 6 bags at a cost of £42.

11 0.00873 m³

12 500 000 cm³

13 No, the total capacity/volume of the tank is 2 090 729.911 cm³ ($\pi \times 55^2 \times 220$), 85% of the capacity is 1 777 120.4 cm³= 1777.1 litres. This will cost £799.70.

14 8

15 515.2 g

16 670 ml

17 7.38 cm

18 7.82 cm

19 19.3 cm

20 3.72 cm

Chapter 22

Check-in questions

1 a Reflection in the x-axis

b Reflection in the y-axis

c Translation by vector $\begin{pmatrix} -7 \\ 0 \end{pmatrix}$

d Translation by vector $\begin{pmatrix} 5 \\ -1 \end{pmatrix}$

2 ab

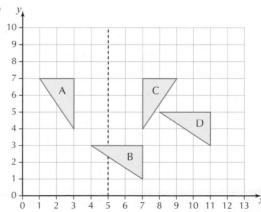

c reflection in $y = x$

3

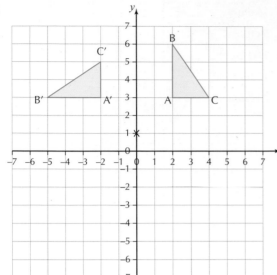

4

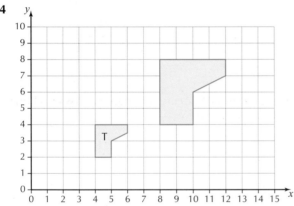

22.1 Translations

1 a 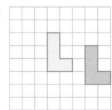 **b**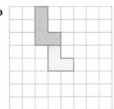

2 a 1 square right, 4 squares up
  b 4 squares left, 3 squares up
  c 5 squares left, 1 square down
  d 1 square left, 4 squares down

3 $\begin{pmatrix} 6 \\ 3 \end{pmatrix}$

4 a $\begin{pmatrix} 3 \\ 3 \end{pmatrix}$ **b** $\begin{pmatrix} 3 \\ 0 \end{pmatrix}$ **c** $\begin{pmatrix} 6 \\ 3 \end{pmatrix}$ **d** $\begin{pmatrix} -6 \\ -3 \end{pmatrix}$

  e $\begin{pmatrix} 8 \\ -2 \end{pmatrix}$ **f** $\begin{pmatrix} 3 \\ -2 \end{pmatrix}$ **g** $\begin{pmatrix} -5 \\ 0 \end{pmatrix}$ **h** $\begin{pmatrix} 0 \\ -5 \end{pmatrix}$

l $\begin{pmatrix} -3 \\ -5 \end{pmatrix}$ **j** $\begin{pmatrix} -5 \\ 5 \end{pmatrix}$

5 a $\begin{pmatrix} 2 \\ -3 \end{pmatrix}$ **b** $\begin{pmatrix} -4 \\ 4 \end{pmatrix}$ **c** $\begin{pmatrix} -4 \\ -4 \end{pmatrix}$

  d $\begin{pmatrix} -2 \\ 3 \end{pmatrix}$ **e** $\begin{pmatrix} 4 \\ 0 \end{pmatrix}$ **f** $\begin{pmatrix} 0 \\ 4 \end{pmatrix}$

22.2 Reflection

1 abc

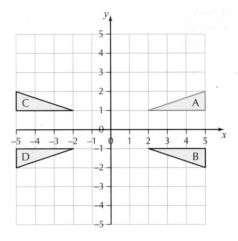

d D is also the reflection of B in the y-axis.

2

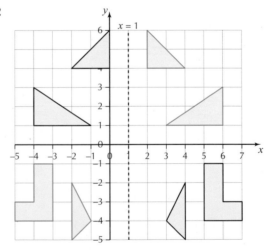

3

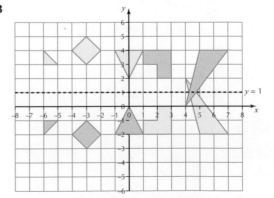

4

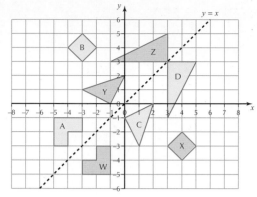

5 a Reflection in the line $x = -3$
  b Reflection in the line $y = -1$
  c Reflection in the line $y = -x$
  d Reflection in the line $y = x$
  e Reflection in the line $y = x$

22.3 Rotations

1 a i D   **ii** A   **iii** D   **iv** D

  b The instruction should include the direction, anticlockwise.

  c Rotation 90° clockwise about O or rotate 270° anticlockwise about O

2 a
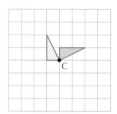
    b
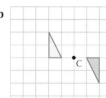

3 a b c

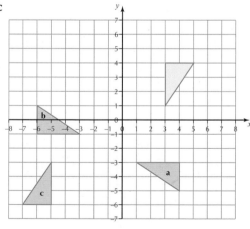

4 a

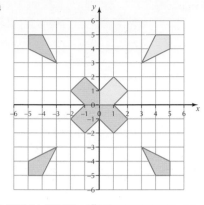

b

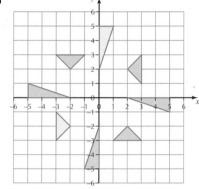

5 Rotation 180° clockwise (or anticlockwise) about the point (4, 3)

22.4 Enlargements

1

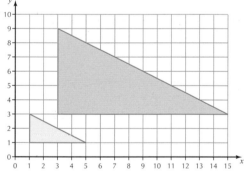

2 a
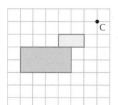
    b

3 a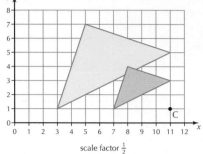

scale factor $\frac{1}{2}$

b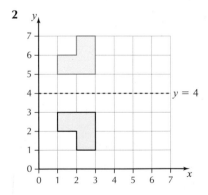

scale factor $\frac{1}{3}$

c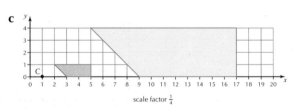

scale factor $\frac{1}{4}$

4 a 20 cm²   **b** 36 cm²   **c** 6 cm²

5 a 3   **b** 8 cm³   **c** 27
  d 27 × 8 = 216 cm³; 6 × 6 × 6 = 216 cm³

Exam-style questions

1 (0, 3), (4, 3), (2, 4), (2, –2)

2

3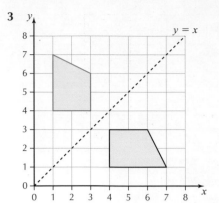

4 a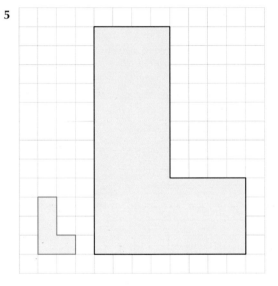

b i true   **ii** false   **iii** true
  iv false   **v** true   **vi** false

5

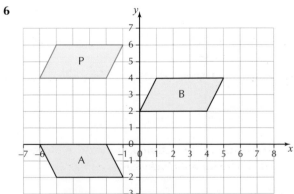

6

7

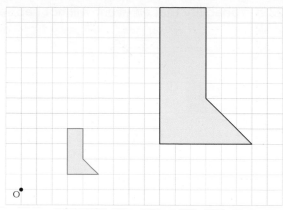

8

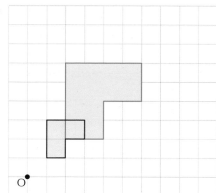

9 a

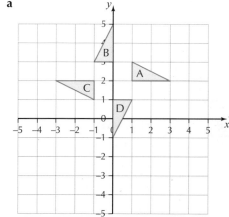

b Rotation 180° anticlockwise (or clockwise) about (0, 2)

10 a Rhombus B is a translation of rhombus A by vector $\begin{pmatrix} 4 \\ -6 \end{pmatrix}$.

b Rhombus B is a rotation of rhombus A of 180° clockwise (or anticlockwise) about (0, 0).

11 a b

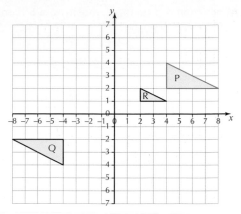

12 24 cm

13 a true   **b** false   **c** true
    d false   **e** false   **f** true

14 a Reflection in the x-axis, or reflection in the line $y = 0$

    b Rotation 90° anticlockwise about (0, 0)

    c

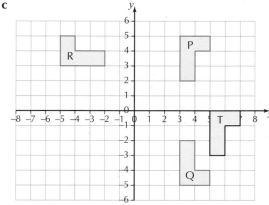

15 a Reflection in the line $y = -x$

    b Rotation 90° clockwise about (–6, 0)

16 a 150 cm²   **b** 337.5 cm²

17 300 cm³

Chapter 23

Check-in questions

1 a very unlikely   **b** impossible   **c** likely

2 a $\frac{6}{11}$ **b** $\frac{5}{11}$ **c** 1   **d** 0

3 0.36

4 $\frac{2}{9}$

5 0.18125

6 Mushroom & pineapple; mushroom & ham or pineapple & ham.

7 a

	1	2	3	4	5	6
1	1	2	3	4	5	6
2	2	4	6	8	10	12
3	3	6	9	12	15	18
4	4	8	12	16	20	24
5	5	10	15	20	25	30
6	6	12	18	24	30	36

b i $\frac{1}{18}$ **ii** $\frac{11}{36}$

8 a $\frac{4}{87}$ **b** $\frac{87}{112}$

9 a

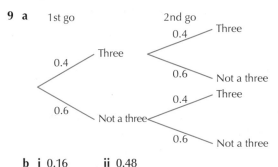

1st go   2nd go

0.4 — Three
0.4 — Three
0.6 — Not a three
0.6 — Not a three
0.4 — Three
0.6 — Not a three

b i 0.16   **ii** 0.48

23.1 The probability scale

1 a $\frac{3}{10}$ **b** $\frac{1}{2}$ **c** 0

2 Probabilities cannot be > 1

3 1

4 These probabilities shown on a scale of 0 to 1:

[**a** $\frac{1}{6}$ **b** $\frac{5}{6}$ **c** 0]

5 Salted, there are more of them.

23.2 Probability of a single event

1 a $\frac{1}{6}$ **b** $\frac{2}{3}$ **c** 0

2 a $\frac{1}{2}$ **b** $\frac{1}{4}$

3 a i $\frac{1}{2}$ **ii** $\frac{1}{4}$ **iii** $\frac{1}{52}$

b Hearts are red and clubs are black, so they cannot both occur on the same card.

4 Student's own response

5 a $\frac{1}{7}$ **b** $\frac{4}{7}$ **c** 0   **d** 1

23.3 Probability that an event will not happen

1 0.8

2 0.95

3

Score	1	2	3	4	5	6
Probability	0.3	0.1	0.04	**0.01**	0.45	0.1

4 $x = 0.1$

5 $\frac{5}{9}$

23.4 Experimental probability

1 500

2 a 100   **b** 200   **c** 250   **d** 0

3 Janet, because her results should be more accurate as she carried out more trials.

4 a

Number of throws	10	50	100	200	500	1000
Number of threes	3	6	19	30	92	163
Experimental probability	0.3	0.12	0.19	0.15	0.184	0.163

b $\frac{1}{6}$ **c** 1000

5 2.8, so 3 days

23.5 Possible outcomes of two or more events

1 a HHH, **HHT, HTH, HTT, THH, TTH, THT,** TTT

b $\frac{1}{8}$ **c** $\frac{3}{8}$

2 a

	1	2	3	4
1	2	3	4	5
2	3	4	5	6
3	4	5	6	7
4	5	6	7	8

b $\frac{3}{16}$

3 $\frac{1}{8}$

4 a 0.27   **b** 850   **c** 357

5 $\frac{1}{18}$

23.6 Tree diagrams

1 a
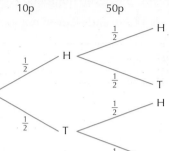

10p   50p

$\frac{1}{2}$ H — $\frac{1}{2}$ H

$\frac{1}{2}$ T

$\frac{1}{2}$ T — $\frac{1}{2}$ H

$\frac{1}{2}$ T

b i $\frac{1}{4}$ **ii** $\frac{1}{4}$ **iii** $\frac{1}{2}$ **iv** $\frac{3}{4}$

2 a
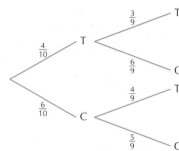

1   2

$\frac{4}{10}$ T — $\frac{3}{9}$ T

$\frac{6}{9}$ C

$\frac{6}{10}$ C — $\frac{4}{9}$ T

$\frac{5}{9}$ C

b i $\frac{2}{15}$ **ii** $\frac{2}{3}$ **iii** $\frac{1}{3}$

3

Prediction   Result

Score
50 — Score — 30 — Score — 24
                        — Not score — 6
  — Not score — 20 — Score — 8
                        — Not score — 12

4 a
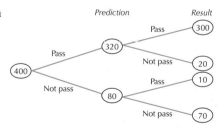

Prediction   Result

Pass
400 — Pass — 320 — Pass — 300
                      — Not pass — 20
  — Not pass — 80 — Pass — 10
                      — Not pass — 70

b 22.5%

5 a
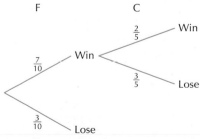

F   C

$\frac{7}{10}$ Win — $\frac{2}{5}$ Win
                 — $\frac{3}{5}$ Lose

$\frac{3}{10}$ Lose

b i $\frac{7}{25}$ **ii** $\frac{41}{50}$

Exam-style questions

1

Strawberry   Honeycomb   Coffee   Fondant or coffee

0 ↓ ↓ ↓ ↓ 1

2 More likely to pick a number card since there are 40 number cards and 12 picture cards.

3 a 35%   **b** 60%

4 0.68

5 0.379

6 a $\frac{1}{3}$ **b** $\frac{8}{9}$ **c** $\frac{2}{3}$

7 No, she is incorrect because the events are independent.

8 0.82

9 0.236

10 0.79

11 No, both have roughly half each for heads and tails.

12 0.32

13 78

14 tea, parkin   coffee, parkin

hot chocolate, parkin

tea, chocolate cake   coffee, chocolate cake

hot chocolate, chocolate cake

tea, apple cake   coffee, apple cake

hot chocolate, apple cake

tea, toffee cake   coffee, toffee cake

hot chocolate, toffee cake.

15 a 6   **b** $\frac{1}{6}$

16 a

	Girl	Boy	total
Vaccinated	48	39	87
Not vaccinated	6	13	19
total	54	52	106

b $\frac{87}{106}$

17 a $\frac{1}{10}$ **b** $\frac{21}{29}$ or 72%

18

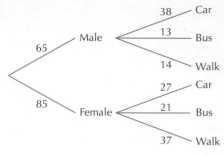

19

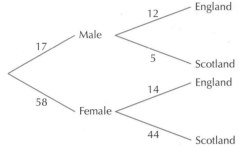

20

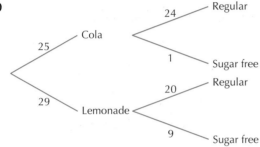

Chapter 24

Check-in questions

1 a {2, 3, 4, 5, 6, 7, 8, 9}   **b** {6, 7}

2 a {1, 2, 3, 5, 7, 9, 11}

b {1, 3, 5, 7, 9}

c {2, 11}

3 a

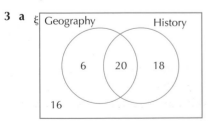

b 18   **c** 26 students   **d** $\frac{4}{15}$

24.1 Set notation

1 A′ → Not A, Ø → The empty set, ∪ → Union of sets, ∩ → The intersection of sets

2 Set B = {A, B, C} and Set C = {A, B, Z}

3 a Set A is a subset of Set B.

4 No, Set D contains only even numbers.

5 a The numbers 1 to 10 and all even numbers greater than 10

b Even numbers > 10

c The numbers 1 to 10

24.2 Venn diagrams

1 G ∩ M

2 b

3

ξ

A — B

Red Green Blue | Yellow

4

ξ

Odd — Square

3
5   1
7   9   4
6   2   8

5 a

ξ

Prime — Factors of 10

3   5   1
7   2   10
8   4   9   6

b i No, there are numbers in the intersection.
ii No, there are numbers outside the circles.

24.3 Using Venn diagrams to solve probability questions

1 5

2 a

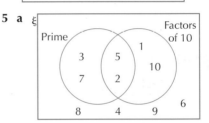

b $\frac{1}{5}$ **c** $\frac{4}{5}$

3 a 41   **b** $\frac{23}{41}$ **c** $\frac{18}{41}$

d $\frac{9}{41}$ **e** $\frac{31}{41}$ **f** $\frac{8}{41}$

4 $\frac{31}{75}$

5 a $\frac{11}{63}$ **b** 3

Exam-style questions

1 Set P = {strawberry}, Set Q = {cherry, banana and custard} and Set T = {toffee, cherry, banana and custard}

2 a {dog, cat, mouse, guinea pig, hamster, tortoise, tiger, lion, panther, leopard, cheetah}

  b {tiger, lion, panther, leopard, cheetah}

  c {dog, mouse, guinea pig, hamster, tortoise}

3 {1, 7, 8, 9, 10}

4 a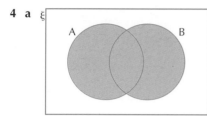

  b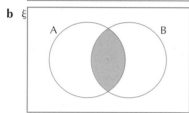

  c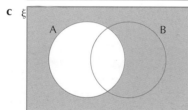

5 a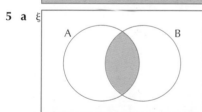

  b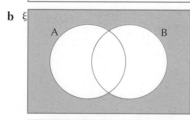

  c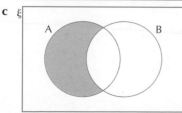

6 c

7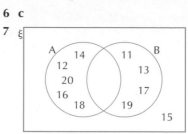

8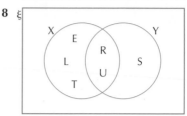

9 Student's own diagram

10 a Student's own diagram   **b** 30

11 7

12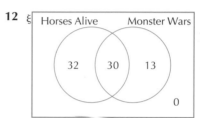

13 a 10   **b** 1   **c** 19

14 $\frac{2}{15}$

15 $\frac{1}{5}$

16 a 27   **b** $\frac{38}{150}$

17 $\frac{12}{82}$

18 a $\frac{12}{50}$ **b** 17

Chapter 25

Check-in questions

1 a Secondary   **b** Primary   **c** Primary
    d Secondary   **e** Secondary

2 For example: February may not be representative of other months.

3

	≤ 20 miles	> 20 miles	Total
Car	45	**55**	100
Train	**120**	**30**	**150**
TOTALS	**165**	85	250

4 a Brisbane   **b** 4 hours

5 a

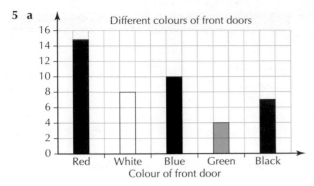

Different colours of front doors

b 44

6 a i Tulips **ii** Daffodils **b i** 2.5% **ii** 7.5%

7 Favourite colour

Green
Blue
Black
Red

8 a 20 **b** 25

25.1 Sampling

1 c Men can run for longer than women

2

	Qualitative	Quantitative	Discrete	Continuous
Colour of car	✓			
Make of car	✓			
Engine size		✓		✓
Number of seats		✓	✓	
Top speed		✓		✓

3 Secondary data, because the event is historic so it's not possible to find the data for yourself.

4 Student's own answer

5 Number each passenger from 1 to 250. Use a random number generator (or tables) to pick the number of people you need, discarding any duplicates or numbers >250

25.2 Organising data

1 Student's own answer

2 The categories overlap. Use inequalities, e.g. $80 < h \le 90$.

3 a 795 **b** 397 **c** 4

4

	0	1	2	3	4	5+	Total
Tablets	3	6	21	24	18	6	78
Phones	1	11	13	44	69	22	160
Total	4	17	34	68	87	28	238

5 Student's own answer

25.3 Diagrams to compare data

1 a 74 **b** 15 **c** Student's own answer

2 a No, 9 sold on Monday and 12 sold on Friday
b Bars are unevenly spaced along the x-axis, scale is not linear

3

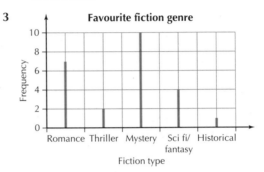

4 Student's own answer

5

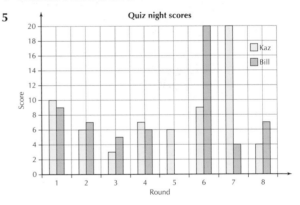

6

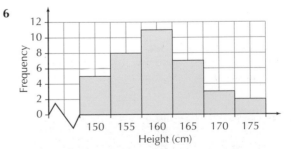

25.4 Pie charts

1

Activity	Number of hours	Angle on a pie chart
Reading	5	30°
Cycling	15	90°
Gym	6	36°
College	24	144°
Work	10	60°

2

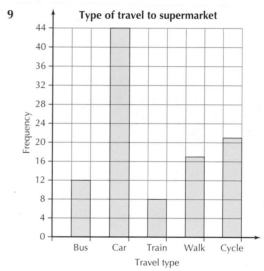

Favourite lecturer

- Mr Fritz
- Mr Barton
- Mrs Bell
- Miss Combwell
- Miss Dhillon

3 a 25%

b Half the people surveyed didn't own any dogs and half the people did own dogs.

c $\frac{1}{6}$ **d** 2400

4 It is impossible to tell from the pie charts alone as we do not know how many people are represented by each pie chart.

5 He can draw them with different sized radii to show they represent different numbers of people or he can write the number of people above each pie chart. This is to avoid confusion in comparing the data.

Exam-style questions

1 Student's own answer

2 Student's own answer

3 a Student's own answer
b Student's own answer

4 The population is biased since people at the gym are likely to use it. Also, would need to collect the data at different times and on different days.

5 16

6

Number of goals	Tally	Frequency
0	ⅣⅢ ‖	7
1	ⅣⅢ ‖	6
2	‖	1
3	‖‖	3
4	‖	2
5	‖	1

7

Data set	Qualitative	Quantitative	Discrete	Continuous
Length of a TV programme		✓		✓
Colour of a shirt	✓			
Dress size			✓	✓
Make of car	✓			

8 a Everyone's opinions are included.

b For example, it's quicker, cheaper, more practical

9

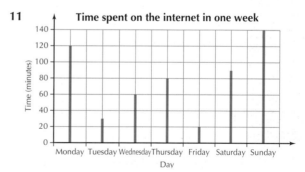

Type of travel to supermarket

10 a 15   **b** 16

c Yes, $\frac{32}{40} = \frac{4}{5}$ which is more than $\frac{3}{4}$

11

Time spent on the internet in one week

12 a 51   **b** 55   **c** 80   **d** $\frac{92}{230} = \frac{2}{5}$

13 a Facebook   **b** 39   **c** 9

14 a 13   **b** $\frac{1}{3}$

15 Tubs of ice cream used last week

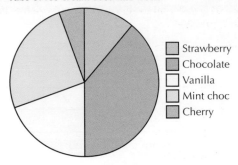

- Strawberry
- Chocolate
- Vanilla
- Mint choc
- Cherry

16 None of the statements are true.

17 Comparative bar charts have the two bars side by side, composite bar charts have them stacked upon each other.

18 You can only represent one set of data on a pie chart. Jessica has two sets of data.

19 a Cars in a supermarket car park

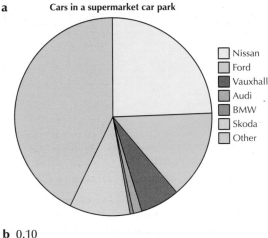

- Nissan
- Ford
- Vauxhall
- Audi
- BMW
- Skoda
- Other

b 0.10

Chapter 26

Check-in questions

1 a 4.75   **b** 9.5

2 a Median 2.5; mode 2   **b** Median 6; mode 6

3 a 6.93   **b** 7   **c** 7

   d 2 is outside the data range

4 150.9

5 a

0	7 9							Key	1	8	= 18
1	8										
2	5 5 5 6 7 7 8										
3	1 1 1 3 6 6 7 9										
4	0 2 2 7 9										
5	0										

b 31

26.1 Averages of discrete data

1 a £29 000   **b** £18 000 and £24 000
   c £24 000   **d** £22 000

2 The mode is the most common value, not the highest value.

3 a 55   **b** no mode   **c** 179.5   **d** 175

4 84 kg

5 He can give the mode (red), because this occurs the most often.

26.2 Finding averages from a frequency table

1 a 3   **b** 2   **c** 4   **d** 3.5

2 a 6   **b** 6   **c** 6   **d** 5.7

3 a 110   **b** 2.6

4 a 12   **b** 8 and 12   **c** 11.5   **d** 11

5 The missing frequency is 10. The mean number of assignments given is 2.35.

26.3 Averages of continuous data

1 a $5 < x \le 10$ **b** 9.4

2 a $20 < x \le 30$ **b** 27
   c The exact values of the maximum and minimum are not known.

3 a $20 < h \le 30$ **b** 25.1   **c** $20 < h \le 30$

4 The exact data values are not known, so it is not possible to calculate the actual mean. The midpoints of the class intervals can be used to calculate an estimate for mean instead.

5 Using the midpoint is more representative and likely to lead to less over- or under-estimating than the minimum or maximum values.

26.4 Stem and leaf diagrams

1

0	00 18 67 79
1	63
2	03 34
3	16 84
4	22
5	69

Key 1 | 63 = £1.63

2 A value that doesn't fit well with the others.

3

13	2
14	1 4 8 9 9
15	0 5 5 7 7 7
16	0 4 4 5 6
17	7 8 8 9
18	0 4 7 7
19	5
20	1

Key 13 | 2 = 132 cm

4 a

2	77
3	33 54 88
4	08 21 79
5	07 11 16 71
6	40

Key 2 | 77 means £2.77

b £4.50   **c** £3.63

5 a

adults		children
	0	6 8 9
2	1	0 2 3 3 6 9
	2	0 2
3	3	
3 2	4	
5 1	5	
4 0 0	6	
6	7	
7 3	8	
14	2	

Key 0|6 means 6 children

b 142 children, it could be the result of a school outing.

Exam-style questions

1 a 2   **b** 8   **c** 2.3

2 7.3 kg

3 68 minutes

4 a 33 cm   **b** 42 cm   **c** 47 cm

5 a 35   **b** 20.6 cm

6 a 24 g   **b** 46 g   **c** 45.6 g

7 No, she needs to put them in order first.

8 The mean value supports the claim.
(497 + 501 + 488 + 504 + 498 + 530 + 487 + 488 + 497 + 510) ÷ 10 = 500

9 a £1600   **b** £1600   **c** £2056.25

10 a

6	9 9 9
7	4 5 6 6 7 7 8 8 9
8	2 3 4 5 7 8
9	0 1 3 4 8
10	0

Key 6 | 9 means 69 years old

b 31 years   **c** 80.5 years

11 a

	Car	Van	Lorry	Bus	Total
Red	2	3	5	7	17
Blue	15	3	9	2	29
White	9	4	0	1	14
Green	2	0	1	0	3
Total	28	10	15	10	63

b blue

12 a 3 minutes   **b** 11 minutes
c The mode for Ruth is less so, on average, she is less late than Josh. However, the range for Ruth is greater, meaning the pattern of her lateness is more erratic.

13 a 6 words   **b** 14.79 words
c The mean for the article on fire safety is higher showing a longer average sentence length. The range is lower for the article on homelessness so the sentence length is less consistent in the article firework safety.

14 a

Jess		Tess
9 9 9 8 8 8 7 7	7	5 6 6 8 8 9
4 1 0	8	0 1 2 4
	9	1

b The median time for both girls is 79 seconds. Jess had Tess have the same average time.

15 7, 12 and 15 must be in the 1st, 3rd and 5th positions. Check student's answers for the other two values.

16 1

17 His score is 48 because it does not affect the mean.

18 The mean uses all of the values. The median is not affected by outliers.

19 $3x + \dfrac{2}{3}$

20 a 171   **b** 80.96

Chapter 27

Check-in questions

1 a

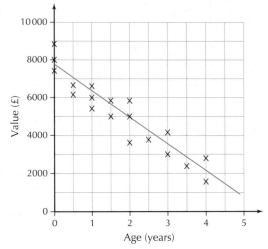

b 2 years   **c** £2900   **d** No

2 a

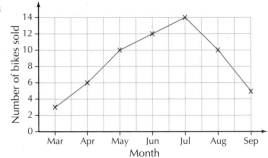

b Sales increase between March and July, but decline from July to September.

27.1 Scatter graphs and correlation

1 Graph A → Positive correlation, Graph B → No correlation, Graph C → Negative correlation

2 ab

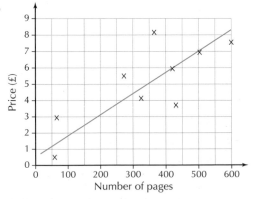

c i Student's own answer ($\pm\frac{1}{2}$ square)

ii Student's own answer ($\pm\frac{1}{2}$ square)

3 a

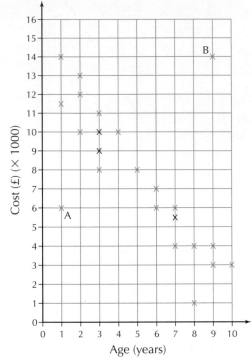

b Yes

c i It's new but has a low value.
ii It's old but has a high value.

d It is not possible to say whether or not this trend will continue as there is no data to make predictions from.

4 a Yes

b Student A had low attendance but a high score in the test

c positive – the greater the attendance, the higher the test result

d Student's own answer

e i Student's own answer - $\pm\frac{1}{2}$ square

ii Student's own answer - $\pm\frac{1}{2}$ square

f This is beyond the range of the data.

5 a

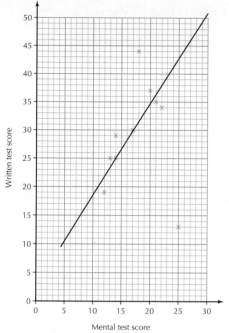

b Student's own answer

c Student's own answer - $\pm \frac{1}{2}$ square

d Dom, because his plot is not near the line of best fit.

e Positive – the higher the mental score, the higher the written score.

27.2 Time series

1 Student's own answer

2 a 17.5, 0000, 0200

b No temperature was recorded at this time.

3 a

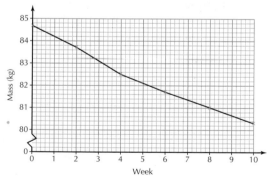

b Student's own answer - $\pm \frac{1}{2}$ square

c Between weeks 2 and 4

d Mass is decreasing; Weight loss is slowing down

e No, we don't have data for this time period and the trend may not continue.

4 a

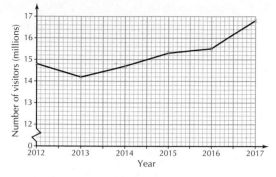

b Student's own answer ($\pm \frac{1}{2}$ square).

c Between 2016 and 2017

d There is an upwards trend, perhaps the result of a good advertising campaign or falling flight prices.

e No, we don't have data that far and the trend may change.

Exam-style questions

1 a

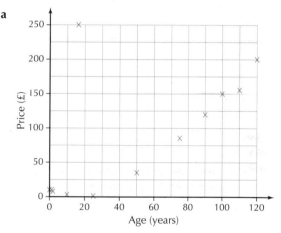

b (16, 25) does not fit the trend

2 a

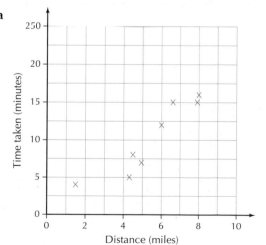

b positive correlation

3 a positive correlation

b the higher the number of visitors, the higher the value of the donations

4 a 3 kg   **b** 47.5 cm

5 a £4.50   **b** 9 km

6 1100 miles

7 The pool was emptied on 8th June.

It rained on June 5th.

The pool was at the normal depth on 3rd June.

8 a

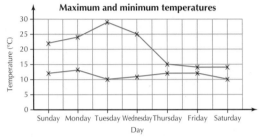

b There are no similarities, but the minimum temperature is fairly constant.

c 2   **d** 0

9 a

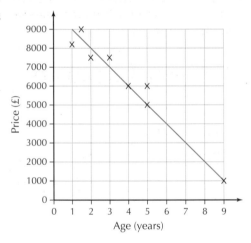

b Student's own answer

c Student's own answer ($\pm \frac{1}{2}$ square)

10 a, b

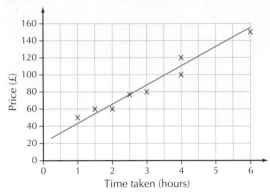

c Student's own answer ($\pm \frac{1}{2}$ square)

d 10 h is outside the data range and extrapolation may be inaccurate.

Formulae you need to learn for your exam

These are the formulae that you need to learn by heart for your Edexcel GCSE (9-1) Maths exam.

Area

Rectangle = $l \times w$

Parallelogram = $b \times h$

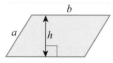

Triangle = $\frac{1}{2} b \times h$

Trapezium = $\frac{1}{2}(a + b)h$

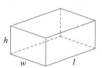

Circles

Circumference = $\pi \times$ diameter, $C = \pi d$

Circumference = $2 \times \pi \times$ radius, $C = 2\pi r$

Area of a circle = $\pi \times$ radius squared, $A = \pi r^2$

Pythagoras

Pythagoras' Theorem

For a right-angled triangle, $a^2 + b^2 = c^2$

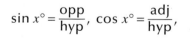

Trigonometric ratios (*new to F*)

$\sin x° = \dfrac{\text{opp}}{\text{hyp}}$, $\cos x° = \dfrac{\text{adj}}{\text{hyp}}$,

$\tan x° = \dfrac{\text{opp}}{\text{adj}}$

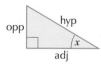

Volumes

Cuboid = $l \times w \times h$

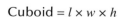

Prism = area of cross section $\times$ length

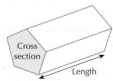

Cylinder = $\pi r^2 h$

Compound measures

Speed

$\text{speed} = \dfrac{\text{distance}}{\text{time}}$

Density

$\text{density} = \dfrac{\text{mass}}{\text{volume}}$

Pressure

The formula for pressure does not need to be learnt, and will be given within the relevant examination questions.

Higher tier formulae

Quadratic equations

The Quadratic Equation

The solutions of $ax^2 + bx + c = 0$,

where $a \neq 0$, are given by $x = \dfrac{-b \pm \sqrt{(b^2 - 4ac)}}{2a}$

Volumes

Volume of pyramid = $\frac{1}{3} \times$ area of base $\times h$

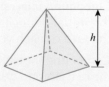

Trigonometric formulae

Sine Rule $\dfrac{a}{\sin A} = \dfrac{b}{\sin B} = \dfrac{c}{\sin C}$

Cosine Rule $a^2 = b^2 + c^2 - 2bc \cos A$

Area of triangle = $\frac{1}{2} ab \sin C$

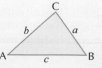

Glossary

Alternate angles – angles formed when two or more lines are cut by a transversal. If the lines are parallel then alternate angles are equal.

Arc – a curve forming part of the circumference of a circle.

Arithmetic sequence – a sequence with a common first difference between consecutive terms.

Bearing – the direction measured clockwise from a fixed point. A bearing has three digits (for angles less than 100°, a zero, or zeros, is placed in front, e.g. 025°).

Bias – a tendency either towards or away from some value.

BIDMAS – a mnemonic that helps you remember the order of operations: Brackets, Indices and roots, Division and Multiplication, Addition and Subtraction.

Centre of enlargement – the point from which the enlargement happens.

Centre of rotation – the point around which a shape can rotate.

Chord – a line joining two points on the circumference that does not pass through the centre of the circle.

Class interval – the width of a class or group, e.g. 0 g, mass of spider < 10 g.

Coefficient – a number or letter multiplying an algebraic term.

Common ratio – the ratio between two numbers in a geometric sequence.

Compound interest – interest that accrues from the initial deposit plus the interest added on at the end of each year.

Compound measure – a measurement using more than one quantity, often using 'per' as in speed, e.g. km/h.

Congruent – exactly alike in shape and size.

Constant of proportionality – the constant value of the ratio of two proportional quantities x and y.

Correlation – the relationship between the numerical values of two variables, e.g. there is a positive correlation between the numbers of shorts sold as temperature increases; there is a negative correlation between the age and the value of cars.

Corresponding angles – angles formed when a transversal cuts across two or more lines. When the lines are parallel, corresponding angles are equal.

Cross-section – the shape of a slice through a solid.

Direct proportion – two values or measurements may vary in direct proportion, i.e. if one increases, then so does the other.

Discrete data – data that can only have certain values in a given range, e.g. number of goals scored, shoe sizes.

Elevation – the 2D view of a 3D shape or object when looking at it from the side or front.

Empty set – a set containing no objects (members).

Enlargement – a transformation of a plane figure or solid object that increases or decreases the size of the figure or object by a scale factor but does not change the shape of the object.

Equation – two terms or expressions connected by an equals sign.

Expression – a statement that uses letters as well as numbers.

Exterior angle – an angle outside a polygon, formed when a side is extended.

Factorisation – finding common factors and putting them outside brackets.

Finite set – a set which has an exact number of members.

Formula – an equation that enables you to convert or find a value using other known values, e.g. area = length × width.

Function – a relationship between two sets of values, such that a value from the first set maps onto a value in the second set.

Geometric sequence – a sequence with a common ratio.

Gradient – the measure of the steepness of a slope:

$$\frac{\text{vertical distance (change in } y)}{\text{horizontal distance (change in } x)}$$ and can be used e.g. to find a derivative such as speed (the gradient) from a distance–time graph.

Highest common factor (HCF) – the highest factor shared by two or more numbers.

Identity – an identity is similar to an equation, but is true for all values of the variable(s); the identity symbol is ≡

e.g. $2(x+3) \equiv 2x+6$

Imperial (units) – traditional/old units of weight and measurements, which have generally been replaced with metric units.

Independent events – two events are independent if the outcome of one event is not affected by the outcome of the other event, e.g. tossing a coin and throwing a dice.

Index (also known as power or exponent) – the small digit to the top right of a number that tells you the number of times a number is multiplied by itself, e.g. 5^4 is $5 \times 5 \times 5 \times 5$; the index is 4.

Inequality – a statement showing two quantities that are not equal.

Infinite set – a set which goes on and on, i.e. it has no end number.

Integer – any whole number, positive or negative, including zero.

Intercept – the point where a line or graph crosses an axis.

Interior angle – an angle between the sides inside a polygon.

Intersection – the point at which two or more lines cross.

Inverse (indirect) proportion – two quantities vary in inverse proportion when, as one quantity increases, the other decreases.

Like terms – algebraic terms that are the same, apart from their numerical coefficients, e.g. $2d$ and $6d$.

Linear sequence – a number pattern which increases (or decreases) by the same amount each time.

Line of best fit – a line (usually straight) drawn through the points of a scatter diagram, showing the trend, which enables you to estimate new values using original information.

Line of symmetry – a line that splits a 2D shape into two equal halves.

Locus (plural: loci) – the locus of a point is the path taken by the point following a rule or rules.

Lower bound – the bottom limit of a rounded number.

Lowest (least) common multiple (LCM) – the lowest number that is a multiple of two or more numbers.

Mean – an average value found by dividing the sum of a set of values by the number of values.

Median – the middle item in an ordered sequence of items.

Metric (units) – units of weight and measure based on a number system in multiples of 10.

Midpoint – the point that divides a line into two equal parts.

Modal class – the largest class in a grouped frequency table.

Mode – the most frequently occurring value in a data set.

Multiplier – the number by which another number is multiplied.

Mutually exclusive events – two or more events that cannot happen at the same time, e.g. throwing a head and throwing a tail with the same toss of a coin are mutually exclusive events.

Net – a surface that can be folded into a solid.

Parallel – lines that stay the same distance apart and never meet.

Percentage increase /decrease – the change in the proportion or rate per 100 parts.

Perpendicular bisector – a line drawn at right angles to the midpoint of a line.

Plan – the 2D view of a 3D shape or object when looking down onto it.

Population – any large group of items being investigated.

Power – the small digit to the top right of a number that tells you the number of times a number is multiplied by itself, e.g. 5^4 is $5 \times 5 \times 5 \times 5$.

Prime factor – a number that is written using only prime numbers multiplied together.

Prime number – a number with only two factors, itself and 1.

Prism – a 3D shape that has a uniform cross-section.

Probability – the chance or likelihood that something will happen. All probabilities lie between 0 (impossible) and 1 (certain). They can be written as fractions, decimals or percentages.

Pythagoras' theorem – the theorem which states that the square on the hypotenuse of a right-angled triangle is equal to the sum of the squares on the other two sides.

Quadratic equation – an equation containing unknowns with maximum power 2, e.g. $y = 2x^2 - 4x + 3$. Quadratic equations can have 0, 1 or 2 solutions.

Quadratic graph – the $\cup$ shaped graph of a quadratic equation.

Random sample – a sampling method in which each data object/person has an equal chance of being selected.

Range – the spread of data; a single value equal to the difference between the greatest and the least values.

Ratio – a way of comparing two or more related quantities. For example, 16 oranges compared to 20 bananas can be expressed as 16 : 20, where the ratio symbol ':' can be read as 'compared to'.

Reciprocal – the reciprocal of any number is 1 divided by the number (the effect of finding the reciprocal of a fraction is to turn it upside down), e.g. the reciprocal of $\frac{2}{3}$ is $\frac{3}{2}$.

Recurring decimal – a decimal that has digits in a repeating pattern, e.g. 0.3333 or 0.252 525.

Reflection – a transformation of a shape to give a mirror image of the original.

Relative frequency –

$$\frac{\text{frequency of a particular outcome}}{\text{total number of trials}}$$

Resultant – the result of adding two or more vectors together.

Roots – in a quadratic equation $ax^2 + bx + c = 0$, the roots are the solutions to the equation.

Rotation – a geometrical transformation in which every point on a figure is turned through the same angle about a given point.

Sample – a section of a population or a group of observations.

Sample space diagram – a probability diagram that contains all possible outcomes of an experiment.

Scalar – a quantity which has only magnitude.

Scale factor – the ratio by which a length or other measurement is increased or decreased.

Scalene – a triangle that has no equal sides or angles.

Scatter diagram – a statistical graph that compares two variables by plotting one value against the other.

Sector – a section of a circle between two radii and an arc.

Set – a collection of objects (members).

Similar – the same shape but a different size.

Simple interest – interest that accrues only from the initial deposit at the start of each year.

Simplify – making something easier to understand, e.g. simplifying an algebraic expression by collecting like terms.

Simultaneous equations – two or more equations that are true at the same time. On a graph the intersection of two lines or curves.

Standard index form (Standard form) – a shorthand way of writing very small or very large numbers; these are given in the form $a \times 10^n$, where a is a number between 1 and 10.

Stem and leaf diagram – a diagram used for displaying data by splitting the values.

Stratified sampling – a sampling method where the population is divided into categories and a sample is taken using the same proportion in each category as in the whole population.

Subset – a set within a set.

Substitution – to exchange or replace, e.g. in a formula.

Supplementary angles – angles that add up to 180°.

Surface area – the area of the surface of a 3D shape, equal to the area of the net of that shape.

Tangent – a straight line that touches a curve or the circumference of a circle at one point only.

Term – in an expression, any of the quantities connected to each other by an addition or subtraction sign; in a sequence, one of the numbers in the sequence.

Terminating decimal – a decimal fraction with a finite number of digits, e.g. 0.75.

Theoretical probability – a predicted probability.

Translation – a transformation in which all points of a plane figure are moved by the same amount and in the same direction. The movement can be described by a vector.

Tree diagram – a way of illustrating probabilities in diagram form. It has branches to show each event.

Trigonometry – the branch of mathematics that shows how to explain and calculate the relationships between the sides and angles of triangles.

Turning point – in a quadratic curve, a turning point is the point where the curve has zero gradient. It could be a minimum or a maximum point.

Universal set – is a set that contains all possible elements.

Upper bound – the top limit of a rounded number.

Vector – a movement on the Cartesian plane described using a column, e.g. $\begin{pmatrix} 3 \\ 4 \end{pmatrix}$.

Venn diagram – a diagram used to represent the relationship between sets.

Vertex – in 2D, a point where two or more lines meet; in 3D, the corners of a shape, where the edges meet.

Index

A

B

BIDMAS

C